About the Author

Alvin Weinberg is one of a band of pioneers whose efforts during World War II culminated in the birth of the nuclear age. He helped design one of the first nuclear reactors, for which he shares a patent, and proposed the first pressurized water reactor, which became the heart of the U.S. Navy's fleet of nuclear submarines. He has served as director of the Oak Ridge Laboratory and as director of the Institute for Energy Analysis at Oak Ridge. He is the author of seven books, including the standard, *The Physical Theory of Nuclear Chain Reactors,* co-authored with Nobelist Eugene Wigner. He is currently a Distinguished Fellow at the Institute for Energy Analysis.

NUCLEAR REACTIONS: SCIENCE AND TRANS-SCIENCE

Masters of Modern Physics

Published Volumes

The Road from Los Alamos by Hans A. Bethe
The Charm of Physics by Sheldon L. Glashow
Citizen Scientist by Frank von Hippel

NUCLEAR REACTIONS: SCIENCE AND TRANS-SCIENCE

ALVIN M. WEINBERG

AIP
The American Institute of Physics

American Institute of Physics
335 East 45th Street
New York, NY 10017-3483

Library of Congress Cataloging-in-Publication Data

Weinberg, Alvin Martin, 1915–
Nuclear reactions : science and trans-science / Alvin M. Weinberg.
p. cm. -- (Masters of modern physics)
Includes bibliographical references and index.
ISBN 0-88318-861-9
1. Science. 2. Science--Social aspects. 3. Nuclear energy.
4. Nuclear energy--Social aspects. I. Title. II. Series.
Q171.W417 1992
303.48'3--dc20 92-16086
CIP

Contents

PART IV. TIME, ENERGY, AND RESOURCES

PART V. NUCLEAR ENERGY

Acknowledgment

Over the years during which I wrote these essays, many people—too many to mention—have helped me clarify my ideas. I would like to thank them, but also to reassure them that I accept responsibility for what I wrote.

I wish to thank Dr. Alfred M. Perry for his help in reading the proof, and Mr. Gunder Hefta for editing the manuscript.

Introduction

I became a nuclear engineer and scientific administrator because of World War II. I had gotten my degree in mathematical biophysics at the University of Chicago in 1939 a few months before Hitler invaded Poland. I had worked with two professors, Carl Eckart and Nicolas Rashevsky; both were theoretical physicists. Rashevsky, the founder of mathematical biophysics, was a marvelously imaginative Russian who saw no reason why he could not create a theoretical structure for biology comparable to the existing structure of theoretical physics; Eckart was a soft-spoken American who had made substantial contributions to quantum mechanics, and later became the world's leading theoretical oceanographer.

Rashevsky's program for mathematizing biology was enormously ambitious: it seemed as if no problem in biology was beyond his grasp. He developed mathematical theories for the division (fission) of living cells, for excitation and transmission of impulses in peripheral nerves, and for the behavior of the central nervous system, as well as for many other biological and sociological phenomena. All of this was, in a way, scientifically premature, and finally I arranged to leave Rashevsky's group to work with the experimental neurophysiologist K. S. Cole. In retrospect, Rashevsky was wonderfully prescient. His ideas about the central nervous system anticipated much of today's analysis both of computers and of nervous systems. His theories of the nerve impulse laid a foundation for work by Alan L. Hodgkin and Andrew F. Huxley in 1952, for which they shared a Nobel Prize (for physiology or medicine) with John C. Eccles in 1963. His mathematical theory of how a living cell divides was isomorphic with the Bohr-Wheeler theory of the fission of uranium!

The announcement in the United States of the discovery of uranium fission by Otto Hahn and Fritz Strassmann coincided almost to the day with my Ph.D. examination in physics and biophysics. All of us at Chicago, as well as everywhere else where physics was done, were electrified by the news. But Rashevsky's group, not knowing *anything* about nuclear physics, was unaware of any connection between mathematical biology and nuclear chain reactions. A year was to pass before the appearance of the Bohr-Wheeler article in which nuclear fission was explained as an instability in the delicate balance between the electrostatic repulsion among the 92 protons in the uranium nucleus and the powerful attractions between the neutrons and protons constituting the nucleus. Gale Young, my lifetime friend and a fellow research associate of Rashevsky's, then published a short note in *Physical Review* in which he pointed out the striking analogy between Rashevsky's dividing cells and Bohr and Wheeler's fissioning uranium nuclei.

I gave little thought to fission until the fall of 1941, when Carl Eckart invited me to work with him, half-time for maybe six months, on the uranium problem. Unbeknownst to me, the nuclear physics group at Chicago under Samuel K. Allison had been investigating the possible role of beryllium in a nuclear chain reaction. Beryllium, when bombarded by a fast neutron, emits two neutrons. Though beryllium by itself can't sustain a chain reaction, it might contribute to a chain reaction based on the fission of uranium. Carl Eckart was trying to calculate the most economical arrangement of beryllium (a very expensive metal) in experiments designed to estimate this "$(n,2n)$" effect, and he needed an eager, but entirely naive, helper like me.

But not quite *entirely* naive. From Rashevsky, I had learned a good deal about the classical mathematical theory of diffusion—in this case, diffusion of nutrients, oxygen, and carbon dioxide into and out of a living cell. Eckart's beryllium experiments involved the diffusion (on a much more gross scale, to be sure) of neutrons: again, a curious coincidental analogy between Rashevsky's mathematics and the mathematics of chain reactors.

So I started to learn about nuclear physics and neutrons, although I regarded the whole thing as an interim activity, to be dropped in favor of serious war work in a few more months.

At the time of the Japanese attack on Pearl Harbor, I had been working seriously on infrared guided missiles and half-seriously on neutrons in beryllium. The Japanese bombs, however, put the Chicago uranium project into high gear, and I was caught up in it. Eckart was to leave Chicago to work in San Diego on undersea warfare. When I lamented his departure, Eckart simply smiled and said, "Wait until Eugene Wigner arrives in Chicago to take my place!"

The Chicago Metallurgical Project

I first met Eugene Wigner in January of 1942. It was early evening when I brought my calculation on beryllium to his temporary office in The University of Chicago's Eckhart Hall. At the time, Wigner was commuting between Princeton and Chicago, before moving to Chicago to head the theoretical group there. I had heard of Wigner's reputation as one of the world's leading theoretical physicists, and I was impressed. I had barely explained what I was trying to do when he frowned for a moment and showed me, without hesitation, what I should have been doing! Indeed, during the almost 50 years that have passed since that first encounter, I have never ceased to marvel that a single individual could have so complete a grasp of physics, chemistry, and mathematics. Somehow, people with this extraordinary breadth don't seem to exist any longer—perhaps because there is so much more to know than there was then.

I became one of Wigner's chief assistants—not because I was the smartest of our group (I wasn't) but because, as Wigner would sometimes put it, he thought I had "human qualities," which he valued, especially when it came to smoothing over the rough episodes that occasionally would crop up at the laboratory. All of which is by way of saying that I fell into administrative work very early in my career, almost without knowing it. (I suppose because there always seemed to be an administrative Alvin Weinberg alongside the technological Alvin Weinberg, I was invited by the American Institute of Physics to collect these somewhat philosophical, somewhat administrative essays.)

Wigner assigned me to be "custodian of the multiplication constant, *k*." Fermi was trying to optimize an arrangement of natural uranium lumps disposed as a lattice in a large block of graphite that would support a self-sustaining nuclear chain reaction. Crucial to this was the multiplication constant, *k*—that is, the number of neutrons produced per neutron absorbed. Unless *k* exceeded 1, no self-sustaining chain reaction was possible. My job was to follow Fermi's experiments, and, applying the theory Wigner had developed, to predict how well an engineered embodiment of various configurations of uranium, graphite, and water (for cooling) would chain-react. For even though Fermi had not yet shown that a chain reaction could be achieved, Wigner and his group were designing the huge water-cooled plutonium-producing reactors that were built at Hanford. (A measure of Wigner's genius is that his basic design of Hanford was completed in January, 1943, only a few weeks after Fermi had demonstrated the first chain reaction, on December 2, 1942.)

Because I was in charge of *k*, I had occasion to examine combinations of uranium and various moderators. Among the first moderators I looked at (with Robert Christy) was ordinary water, which we discovered could sustain a chain reaction only if the uranium were slightly enriched. The idea of pressurized-water reactors using slightly enriched uranium was therefore never very far from my mind—and when, several years later, Hyman Rickover came on the scene, I urged him to adopt the pressurized-water system for his *Nautilus*. That pressurized-water reactors then came to be the dominant type for civilian power came as a surprise to me.

I write these words almost fifty years after these events occurred. Today, nuclear energy, for which we had had so much hope, has foundered. The first nuclear era has ended. Did we, in those earliest days, recognize the possibility of such a denouement? Not really—we were too busy trying to win the war, and making plutonium. To be sure, by 1944 we began to think about the future of nuclear energy, but, except for Fermi, none of us realized that public acceptance could be the Achilles' heel of nuclear energy. Fermi, with his usual caution, spoke to us on one occasion—it must have been spring, 1944—about as follows: "In producing power from fission, we are creating radioactivity on an unprecedented scale—a scale that humankind has had absolutely no experience with. Whether technology that produces such vast amounts of radioactivity will be accepted by the public is an open question."

I don't think Fermi realized how true this rather off-hand remark would turn out to be. As for me, engaged as I have been all my life in the development of this technology, these questions have engaged an ever-increasing part of my energies. I have written many dozens of articles on the social aspect of nuclear energy and have collected earlier writings in two books, *Reflections on Big Science* (MIT Press, 1967), and *Continuing the Nuclear Dialogue* (American Nuclear Society, 1985). In Part V of this volume, I include a small sampling of my writings on nuclear energy; most of them have been written since the publication of *Continuing the Nuclear Dialogue*.

Clinton Laboratories and Oak Ridge

Wigner visualized a large laboratory devoted to development of nuclear energy after the war ended. To this end, he persuaded the Army to convert the wartime pilot plant, Clinton Laboratories in Oak Ridge, Tennessee, into a full-fledged nuclear-energy development laboratory, and he agreed to serve as research director during this transition. Clinton Laboratories eventually became the Oak Ridge National Laboratory (ORNL). When Wigner returned to Princeton in 1947, I was asked to be either the research director or the director by Clark Center, the vice president of Union Carbide, the contractor for the laboratory. At the time, I had illusions about retaining my hand directly in scientific work, and so I chose the title Associate Director—but this was hardly realistic. I became Research Director in 1948 under our new Director, Clarence Larson; and I became Director of the Laboratory in 1955, when Clarence (who later became a U.S. Atomic Energy Commissioner) joined Union Carbide. I remained Director of ORNL until the end of 1973, a period of 18 years.

I always regarded being Director of ORNL as one of the best scientific-administrative jobs in America. It was not that the pay was all that good, or the red tape all that manageable. It was simply that one kept in touch with most of modern science, if not as a participant, then as a highly interested kibitzer. More than that, one met all the best (or, at least, the most important) people, and one had the opportunity to make lots of speeches. I used to get terrible stage fright when I spoke in public, a difficulty I finally overcame simply by giving more than my share of speeches. In any event, this habit of speechifying, and my position as Director of ORNL, seems to have come to the attention of George Kistiakowsky, President Eisenhower's science advisor. In 1959, he invited me to join the President's Science Advisory Committee (PSAC). As a member of PSAC, I embarked on an avocation, if not a second career, as a philosopher of scientific administration. The essays in Parts I and II of this book, "Science and Trans-Science" and "Scientific Administration," were engendered by my three-year tenure on PSAC. Once challenged by these "trans-scientific" questions, I became hooked, and I continue to write and lecture on these subjects. In these endeavors, I owe much to my good friend Edward Shils, the eminent sociologist and editor of *Minerva*, who, during our 30-year friendship, has encouraged me to speak, write, and publish (often in *Minerva*) on the general subject of science and trans-science.

The Institute for Energy Analysis

Already, in 1970, I found my views on nuclear energy increasingly uncongenial to the Atomic Energy Commission. I realized that something was amiss when, at about this time, Congressman Chet Holifield, Chairman of the Joint Committee on Atomic Energy, suggested to me that, if I had misgivings about the safety of reactors, perhaps it was time to leave Oak Ridge National Laboratory. I had also begun to sense, without quite being told, that I was regarded as too impractical by some of the AEC staff. So, by 1973, I was, to put it bluntly, fired as Director of ORNL. I had never been fired from any job before, and this was fairly bitter medicine; but, as things turned out, it was one of the best things that ever happened to me.

Simultaneous with my leaving ORNL came the first Arab oil crisis: energy quickly became the nation's—indeed, the world's—primary problem. At the time, William O. Baker was a sort of scientific grey eminence in the Nixon White House. He and I had been good friends ever since we served together on PSAC; and he encouraged me to organize an Institute for Energy Analysis (IEA)—a think-tank, to be funded by the U.S. government, and whose purpose, broadly, was to formulate coherent energy policy for the United States.

With the help of Baker and Dixy Lee Ray—who, by this time, had succeeded James Schlesinger as Chairman of the AEC—we organized IEA as a component of the Oak Ridge Associated Universities. We planned to open for business on January 1, 1974, but, late in December, I received a call from John Sawhill, a member of the White House Staff at the time, who told me "We want you to head the Energy Research and Development Office in the White House."

I am forever grateful to H. G. MacPherson for taking over IEA as Acting Director: on his shoulders fell the onerous task of getting IEA started, which he did with great distinction.

So I spent the next 347 days in Washington heading a small energy research group in the Old Executive Office Building. Did we accomplish much during this frantic year? It is very hard to say. Perhaps two tangible things came out of our office—the establishment of the Solar Energy Research Institute and the conversion of AEC into the Energy Research and Development Administration, which later became the Department of Energy.

The year 1975 saw me back at IEA, from which I retired in 1985. The essays in Part IV of this book, "Time, Energy, and Resources," stem from this period. As I think about these matters, I regard our

Institute's emphasis on time as a key but insufficiently recognized element of energy policy as centrally important. Some of my fellow physicists invented second-law efficiency—the idea that the quality of a heat source should match the quality of the job the heat source is called upon to perform. Unfortunately, this maxim ignores time; if a heat source and heat sink match perfectly, the job requires infinite time to accomplish! I regard this as the single most important idea IEA contributed to the energy debate: it was formulated by Daniel Spreng and, independently, by Pim Van Gool.

Strategic Defenses and Arms Control

Eugene Wigner had, from the very beginning of the nuclear era, been preoccupied with ways to maintain both peace and freedom in a thermonuclear world. He therefore devoted much of his later life to devising defensive strategies—including civil defense, though he recognized, of course, that no defensive strategy could be perfect in a world of 30,000 thermonuclear warheads. I invited him to ORNL to set up a group to study civil defense; and my somewhat unorthodox view on the role of defensive systems in a nuclear world have been much influenced by the doctrines developed by the ORNL civil-defense group. I turned to a serious re-examination of strategic defense and arms control in 1983, following President Reagan's famous Star Wars speech. My views on these matters, largely formulated in collaboration with my colleague at IEA, Dr. Jack Barkenbus, constitute Part III of this collection, "Strategic Defense and Arms Control."

In Retrospect

People sometimes offer their sympathy to me because the 50 years of my scientific career, which began so excitingly with nuclear energy as the great hope of mankind, is closing with nuclear energy being rejected by most of the public. Has my life been spent in vain, as we nuclear Prometheuses come to no better end than the original Prometheus?

Heavens no! In the first place, I've had a marvelous time—there has almost never been a boring moment. Few scientific administrators can match the excitement, the excellence (as Robert Oppenheimer put it) of a life devoted to helping create the first nuclear era.

And although I am largely retired now, I remain highly optimistic about the future and about nuclear energy. That the bomb gave us the time needed to convert aggressive Marxism-Leninism into something less threatening is now obvious. And the new nuclear-power systems, with their much higher levels of safety, I have little doubt will provide the technical basis for a reborn, second nuclear era—one that will be graced by an unaccustomed peace.

I suppose my main regret is that, being 77 years old, I will not be around to see the second nuclear era. But there will be many enthusiastic, younger nuclear engineers—perhaps even some who converted from biophysics to nuclear energy—who will shape a better second nuclear era and who will be around to enjoy the fruits of their endeavors.

Part I: Science and Trans-Science

That science cannot solve every problem encountered by humankind is obvious. Yet, as our society becomes increasingly scientific in outlook, we expect science to give us unequivocal answers where in principle such answers are impossible to obtain. This dilemma probably was first recognized in the case of nuclear energy. Here was a technology that sprang full-blown from science—but the many controversies that nuclear power spawned too often involved questions that could be posed in a scientific idiom yet could not be answered by science. I suppose it was my participation, from the very beginning, in the nuclear enterprise that caused me to realize that many of the issues that evoke the bitterest debate depend upon disagreements about the interpretation of data at the very limit of science. This led me in 1972 to write the essay "Science and Trans-Science."

The second essay, "The Regulator's Dilemma," written in 1985, brings the 1972 essay up-to-date.

In 1966, at the height of my nuclear euphoria, I visualized nuclear energy—and, by implication, other technological marvels—as being magical panaceas for much of what troubled us. From today's (1991) standpoint, my optimism about nuclear energy seems utopian; however, the basic idea of technological fixes still is sound. I coined the phrase "technological fix" at the time, not realizing that exactly the same idea had been put forward by Richard L. Meier in his 1956 book *Science and Economic Development: New Patterns of Living* (MIT Press). At the time I wrote "Can Technology Replace Social Engineering?" I did not quite realize that, at least in principle, a perfected technology could quiet some of the most blatant criticisms of technology. As the nuclear debate escalated, I gradually recognized this possibility, as I describe in Part V of this volume.

Science and Trans-Science

Much has been written about the responsibility of the scientist in resolving conflicts that arise from the interaction between science and society. Ordinarily, the assumption is made that a particular issue on which scientific knowledge is drawn into the resolution of a political conflict—for example, whether or not to build a supersonic transport (SST) or whether or not to proceed with a trip to the Moon—can be neatly divided into two clearly separable elements, one scientific, the other political. The scientist is expected to say whether a trip to the Moon is feasible or whether the SST will cause additional skin cancer. The politician, or some other representative of society, is expected to say whether the society ought to proceed in one direction or another. The scientist and science provide the means; the politician and politics decide the ends.

This view of the role of the scientist and, indeed, of science itself, is, of course, oversimplified, in particular because, even where there are clear scientific answers to the scientific questions involved in a public issue, ends and means are hardly separable. What is thought to be a political or social end turns out to have numerous repercussions, the analysis of which must fall into the legitimate jurisdiction of the scientist, and each of these repercussions must also be assessed in moral and political terms; or what is thought to be a scientific means has nonscientific implications, which also must be assessed in these terms. The relationship between the scientist and the politician is far more complicated than the simple model described above.

Many of the issues that arise in the course of the interaction between science or technology and society—the deleterious side effects of tech-

First published in Minerva **10**, pp. 209–222 (1972). Reprinted by permission.

nology, or the attempts to deal with social problems through the procedures of science—hang on the answers to questions that can be asked of science and yet *cannot be answered by science.* I propose the term *trans-scientific* for these questions since, though they are, epistemologically speaking, questions of fact and can be stated in the language of science, they are unanswerable by science; they transcend science. Insofar as public policy involves trans-scientific rather than scientific issues, the role of the scientist in contributing to the promulgation of such policy must be different from his role when the issues can be unambiguously answered by science. I shall examine this role of the scientist, particularly the problems that arise when scientists can offer only trans-scientific answers to questions of public policy in situations in which the public, politicians, civic leaders, etc., look to scientists to provide scientific answers.

Examples of Trans-Scientific Questions

Biological effects of low-level radiation insults

Consider the biological effects of low-level radiation insults to the environment—in particular, the genetic effects of low levels of radiation on mice. Experiments performed at high radiation levels show that the dose required to double the spontaneous mutation rate in mice is about 100 roentgens of x rays. If the genetic response to x radiation is linear, then a dose of 150 millirems would increase the spontaneous mutation rate in mice by 0.15 percent. This is a matter of importance to public policy, since the various standard-setting bodies had decided that a yearly dose of about 150 millirems (actually 170 millirems) to a suitably chosen segment of the population was acceptable. Now, to determine, at the 95 percent confidence level by a direct experiment, whether 150 millirems will increase the mutation rate by 0.15 percent requires many billions of mice! Of course, this number falls if one reduces the confidence level; at the 60 percent confidence level, the number is reduced 40-fold. Nevertheless, the number is so staggeringly large that, as a practical matter, the question is unanswerable by direct scientific investigation.*

---

*To be sure, indirect evidence about the shape of the dose-response curve for x rays at very low dosage can be inferred from experiments that measure the relative biological effectiveness of highly ionizing radiation and x rays. Such experiments suggest that the dose-response curve for x rays at low dosage is quadratic, not linear. However, these experiments are suggestive, not definitive: they still represent extrapolations to very low doses of radiation of the observations taken at high dose.

Of course one may ignore dose-rate effects and increase the dose 30-fold to 5 rem in order to

This kind of dilemma is not confined to radiation. No matter what the environmental insult, to measure an effect at extremely low levels usually requires impossibly large protocols. Moreover, no matter how large the experiment, even if *no* effect is observed, one can still only say there is a certain probability that in fact there is no effect. One can never, with any finite experiment, prove that any environmental factor is totally harmless. This elementary point has unfortunately been lost in much of the public discussion of environmental hazards.

The probability of extremely improbable events

Another trans-scientific question is the probability of extremely unlikely events—for example, a catastrophic reactor accident, or a devastating earthquake that would, say, destroy Hoover Dam and thereby wash out parts of the Imperial Valley of California. Probabilities of such events are sometimes calculated. For example, in the case of a catastrophic reactor accident, one constructs plausible accident trees, each branch of which is triggered by the failure of a particular component. Statistics about the reliability of each component are often known, since many components of the type under consideration—ion chambers, transistors, control-rod bearings—have been tested. But the calculations are obviously suspect: first, because the total probability obtained by such estimates is so small—say 10^{-5}/reactor/year—and, second, because there is no proof that every conceivable mode of failure has been identified. Because the probability is so small, there is no practical possibility of determining this failure rate directly—by building, let us say, 1,000 reactors, operating them for 10,000 years, and tabulating their operating histories.

These two examples illustrate questions that are trans-scientific because, although they could conceivably be answered according to strict scientific canons if enough time and money were spent on them, to do so would be impractical.*

---

simulate the possible life dose to a woman during her child-bearing period. Even in this case, at the 95 percent level, about 10 million mice would be required. But, in any event, there are serious difficulties in extrapolating to man data found in mouse experiments.

*After this article was published, Harvey Brooks pointed out that for systems governed by nonlinear equations, tiny changes in the initial conditions can lead to very large changes in the ultimate behavior of the system—and indeed can lead to chaos. The behavior of such systems is indeterminate ("trans-scientific") since initial conditions can never be established with absolute precision. H. Poincaré gave a mathematical analysis of such systems.

Engineering as trans-science

Engineering, especially in fields that are developing rapidly, typically requires decisions made on the basis of incomplete data. The engineer works against rigid time schedules and with a well-defined budget. He cannot afford the luxury of examining every question to the degree that scientific rigor would demand. Indeed, "engineering judgment" connotes the ability, as well as the necessity, to come to good decisions with whatever scientific data are at hand. Sometimes the crucial data are insufficient for the engineer to proceed: the project then must await further scientific research. Usually, however, the engineer makes do with whatever data he has: he then uses the wisdom called "engineering judgment" as a guide.

The engineer exercises his judgment, on the whole, by being conservative. If he is unsure of the "creep" behavior of a new alloy, he will ordinarily overdesign his sections so as to withstand the worst conditions he can imagine. The extent of overdesign is largely determined by the engineer's budget: an important incentive for acquiring more data is the desire to avoid costly overdesign.

Uncertainty is, in a sense, inherent in engineering: unless one is willing to build a full-scale prototype and to test it under the precise conditions that will be encountered in practice, there is always the uncertainty of extrapolating to new and untried circumstances. Where the device being engineered is small, like a jet engine, a full-scale prototype is customarily built: difficulties are worked out either on the prototype or on the early production models. But where the device is huge—like the Aswan Dam, or a 1,000-MW plutonium breeder, or a large bridge—a full-scale prototype is out of the question. Moreover, the service life of such devices may be as long as 100 years: even if a prototype were built, there would be little sense in waiting until weaknesses appeared in the prototype before starting on the next model. Thus in every advancing technology there are inherent elements of scientific uncertainty that, as a matter of principle, can never be totally resolved. In this sense, such technologies are trans-scientific, or at least possess trans-scientific elements. And, indeed, most of the examples I use in the remainder of my discussion are derived from technology, particularly the technology of nuclear reactors.

Trans-scientific questions in the social sciences

In the social sciences, trans-scientific questions arise very frequently. One often hears social scientists classify questions as being "researchable" or "not researchable." In the former category are, presumably, questions that, at least in the estimate of the social scientist, can be approached with some hope of success. In the latter category are those that cannot.

What makes a question in the social sciences resistant to investigation or trans-scientific? Before the advent of the large computer, I suppose many problems in social science entailed operations that were beyond the capacity of available personnel. Obviously, the computer has changed this. But there remains a very important class of seemingly social and scientific questions that will always be in the realm of trans-science.

I refer to the behavior of a particular individual. In physics, if we know the initial position and velocity of a specific macroscopic object and the forces acting upon it, we can predict its trajectory—not the trajectories on the average of many objects like this one but the trajectory of this particular object. The physical sciences are capable of predicting particular macroscopic events precisely from the laws of nature and from the initial conditions. Even in quantum physics, we can make precise predictions of the behavior of a collection of atoms or molecules, and the statistical distribution of the behavior of the microscopic identities. This enormous proficiency is attributed by Elsasser to the homogeneity of the class of objects of discourse in quantum physics—every hydrogen atom is the same as every other hydrogen atom, and statistical variability can itself be predicted. In contrast, the social sciences deal with classes, the individual members of which display wide variability as well as being subject to the vagaries of consciousness. The predictions of social sciences are inevitably less reliable than are those of the physical sciences. Moreover, insofar as the social sciences can predict behavior, it is the behavior, *on the average*, of large classes. To expect the social sciences to predict individual behavior, or even individual events, with anything like the precision we expect of the physical sciences is generally to ask too much. Yet public policy often requires estimates of the future behavior of individuals—for example, in the Cuban missile crisis, President Kennedy had to make some estimate of the behavior of Premier Khrushchev. Even where many individuals are involved and the event to be predicted is a unique constellation of the action of many people, the proficiency of the social sciences is less than that of the physical sciences. Nor is this simply a matter of the social sciences being "less well developed" than the physical sciences; it is my impression that there are basic limitations to the predictive powers of the social sciences that derive from the inherent variability and consciousness of the individuals who make up the populations studied by social science. From this point of view, one would argue that much of social science (perhaps with the exception of economics) is indeed trans-science: that its proficiency in *predicting* human behavior is, and probably always will be, far more limited than that of the natural sciences.

Axiology of science as trans-science

Still a third class of trans-scientific questions constitutes what I call the axiology of science; these are questions of "scientific value," which include the problem of establishing priorities within science. These are the problems discussed under the name of criteria for scientific choice, as well as the valuation of different styles of science: pure versus applied, general versus particular, spectroscopy versus paradigm-breaking, search versus codification. All of these are matters of "scientific values" or taste rather than scientific truth. Insofar as value judgments—that is, ultimate questions of *why* rather than proximate questions of *what*—can never be answered within the same universe of discourse as the one in which the question arose, any resolution of these issues clearly transcends science, even though the issues themselves seem to be internal to science.

The examples I have offered transcend science in three rather different senses. In the first case (low-level insult), science is inadequate simply because to get answers would be impractically expensive. In the second case (social sciences), science is inadequate because the subject matter is too variable to allow rationalization according to the strict scientific canons established within the natural sciences. And in the third case (choice in science), science is inadequate simply because the issues themselves involve moral and aesthetic judgments: they deal not with what is true but rather with what is valuable.

Trans-Science and Public Policy

Increasingly, society is required to weigh the benefits of new technology against its risks. In such a balance, both scientific and trans-scientific questions must be asked by those who have the responsibility for the decisions and those who concern themselves with the decisions. The strictly scientific issues—whether, say, a rocket engine with enough thrust to support a manned moon shot can be built—can in principle be settled by the usual institutional mechanisms of science, such as debate among the experts and critical review by peers. But what about the issues that go beyond science, on which the scientist has opinions that, however, do not carry the same weight as his opinions that are based on rigorous scientific evidence? These issues are dealt with by two institutional mechanisms: the ordinary political process and adversary procedures.

The political process, in a general sense, establishes priorities: it allocates scarce resources among alternative uses where there is no marketplace and where there is no objective, norm, or standard to govern the allocation. The resources to be allocated may be tangible and specific, as when a decision is made to go to the Moon; or they may be much more diffuse and pervasive, as when a national commitment is made to improve the position of minorities. In either case, the resources are allocated and the priorities are established by the interplay of competing political views and powers: those who want to build the SST exert what political power they have—the capacity to summon votes, to grant favors, to threaten to withdraw support—and this is resisted by those who dislike SSTs, for whatever reason. In such specific allocations of scientific resources, the scholarly discussion of science policy, dealing as it does with matters that are not internal to science, is intended to elevate and illuminate the political discussion, at whatever level this occurs. It seeks to make the contestants in the conflict more aware of the consequences of any decision and of alternatives, to show them what its implications are in terms of other values, to ensure that they weigh the costs of what they seek, and that they are aware of the values that are implied in their choices.

In the other, more subtle and pervasive working of the political process—the establishment of social priorities—scientific elements are sometimes involved. To take the case of the position of blacks in the United States, the Supreme Court invoked a "scientific" doctrine—that educational deprivation caused psychological damage to the individual—as an important argument for its decision to order desegregation. And in the political processes that have followed this decision, this finding from the sphere of social psychology has never been far from the political debate.

The other institutional mechanism for arriving at decisions is the adversary procedure. Although adversary procedure of a sort is implicit in any confrontation, public or private, I shall reserve the term for those formal, legal, or quasi-legal proceedings at which proponents, both scientists and nonscientists, of opposing views are heard before a body or an individual who is empowered to render a decision after having heard the conflicting contentions. Before a permit is granted for the construction of a nuclear reactor in the United States, for example, the applicant must receive a licence from a licensing board. The board must find that the reactor can be operated "with reasonable assurance that the health and safety of the public is maintained."

The hearings before the board have a legal aspect. Those who oppose granting a license, usually because they disagree with the likely answers to the questions about the safety of nuclear reactors, appear as interveners. The procedure pits one adversary against another. The arguments used bear close resemblance to those used in the political process, but they are generally more factual because there is face-to-face confrontation and opportunity for cross-examination; there are certain procedural rules of rhetorical etiquette that inhibit the demagogy and exaggeration common in the political process; the contending parties are also usually better informed than impassioned participants in the political process; and, finally, they know that they will be asked specific questions by the board or some quasi-judicial equivalent. Through this confrontation, trans-scientific and scientific questions related to the side effects of nuclear reactors are resolved.

The adversary procedure is likely to be used increasingly in modern, liberal societies in their attempts to weigh the benefits and risks of modern technology. Certainly this is the case in the United States. The United States Environmental Protection Agency, for example, requires statements regarding "environmental quality" from the promoters of any large technological enterprise that might affect the environment. These statements, if challenged, undoubtedly will lead to lengthy and elaborate adversary proceedings.

It is therefore important to examine the validity of formal adversary procedures for settling technological or semitechnological issues. Harold P. Green has argued that, in adversary procedures, representatives of the public are usually less well informed than representatives of the applicant and that, therefore, the former are at a disadvantage. This places a heavy responsibility on the agency before which such adjudicative procedures are held to try to redress any such inequality in the positions of the contending parties. To a great extent, this now happens in the United States Atomic Energy Commission's review of nuclear reactors. The regulatory staff of the commission subject every application for a nuclear reactor to a searching and highly informed technical scrutiny: the public adversary procedure is the culmination of a lengthy prior analysis by the staff of the Atomic Energy Commission. Professor Green asserts that the regulatory staff of the commission, at such hearings, join with the applicant against the public interveners.[1] This is hardly the view of many applicants who are often distressed and frustrated by the painstaking and slow course that these reviews require.

Whether the adversary procedure is adequate or not seems to me to depend on whether the question at issue is scientific or trans-scientific. If the question is unambiguously scientific, then the procedures of science, rather than the procedures of law, are required for arriving at the truth. Where the questions raised cannot be answered from existing scientific knowledge or from research that could be carried out reasonably rapidly and without disproportionate expense, then the answers must be trans-scientific, and the adversary procedure seems therefore to be the best alternative. In principle, one exhausts all of the *scientific* elements, one answers every question that can be answered scientifically, before dealing with the trans-scientific residue. With regard to the public hazard of the SST, the scientific evidence for the connection between increased sunlight and skin cancer seems to me to be unequivocal, and I believe experts agree; the matter can be settled by the usual institutional procedures of science. The effect of nitric oxide exhaust from SST engines on the ozone concentration in the stratosphere has less direct empirical evidence to support it and therefore is more controversial: this part of the issue contains both scientific and trans-scientific elements and might be illuminated by adversary procedures. Finally, the question of whether or not to go ahead with the SST with the evidence at hand is an issue that is primarily nonscientific—for example, the cost as compared to a wide variety of other competing activities. This must be decided by political processes because ultimately the decision to proceed or to desist is a matter of ethical or aesthetic values. Where there is no consensus on these values, the process of decision must be political.

To one trained in the law, a rather formal and somewhat stylized adversary procedure might seem to be a reasonable institutional arrangement for arriving at truth—whether it be legal, trans-scientific, or scientific. But to the scientist, adversary procedures seem inappropriate and alien. To be sure, such procedures are useful in establishing the credibility of witnesses—that is, in establishing whether the witness is stating the whole truth and nothing but the truth. In science, however, the issue is not credibility but specific competence—that is, the witness's ability to recognize and know scientific truth—and this is not reliably established by an adversary procedure conducted by lawyers rather than scientists. On the other hand, in trans-science, where matters of opinion, not fact, are the issue, credibility is at least as important as competence. One must establish what the limits of scientific fact really are, where science ends and trans-science begins. This often requires the kind of selfless honesty that a scientist or engineer with a position or status to maintain finds hard to exercise. For example, in the acrimonious debate over low-level radiation insult between J. W. Gofman and A. R. Tamplin, on the one hand, and most of the nuclear scientists on the other, neither side was quite willing to say that the question was simply unresolvable, that this was really a trans-scientific question. The adversary procedure undoubtedly has considerable merit in forcing scientists to be more honest, to say where science ends and trans-science begins, as well as to help weigh ethical issues that underlie whatever choices society makes between technological alternatives.

There is yet another possible way to resolve some of the unanswerable questions of public or environmental risk caused by new technology: this is to perfect the technology so as to minimize the risk. We say that there is a possibility (which we cannot quantify) that low-level radiation insult will cause cancer. We can never eliminate these insults entirely—our technology is too necessary for our survival to dismantle, and it is idle to hope that we shall ever have technology with absolutely no risk.

To be sure, we shall always try, through improved technology, to reduce effluents and other by-products from any device. In some measure, this is how the debate over the radioactive emission standards from nuclear reactors is being resolved. The original Atomic Energy Commission regulations permitted doses of up to an average of 170 millirems per year to groups of individuals in the vicinity of a nuclear installation. The controversy over these standards has been resolved by technology: nuclear reactors now can be built that emit only 1/100 or less of the original standards, and the standards are being lowered accordingly. The question of the permissible dose has been moved from the sphere of trans-science toward that of science—emissions have now been moved much closer to the point (zero emission) where all scientists can agree by scientific standards that there is no danger.

Even the residual risk, the magnitude of which cannot really be determined by science, can be reduced if science can develop a cure for the untoward biological side effects of the environmental insult. This argument has been put forward by H. I. Adler of the Oak Ridge National Laboratory, and I believe it deserves serious consideration. Suppose we developed a safe and simple method of immunizing against cancer. That this is no longer a fantasy is believed at least by the panel that advised United States Senators Yarborough and Kennedy to launch a new cancer program. Attitudes toward residual and unavoidable contamination of the environment would certainly be modified if there were some form of immunization against the side effects that gave rise to concern in the first place.

The possibility of genetic intervention would also help to eliminate the issue of residual contamination of the biosphere. At present, there are 30 or more enzyme deficiencies, presumably of genetic origin, that can be detected in amniotic fluid. If science could, by amniotic analysis and therapeutic abortion, reduce the risk of genetic abnormality from whatever cause by a large factor, our attitude toward the trans-scientific question of low-level radiation insult would be significantly affected. In offering this possibility, I realize that abortion, for whatever reason, raises grave moral and social questions. To me the moral scale weighs heavily against bringing into the world babies who are predestined to short lives of torture; to others, the balance may come out differently.

The Republic of Trans-Science and the Political Republic

The validity of scientific knowledge is established and maintained through the critical judgement of scientific peers. The whole system is described by Michael Polanyi very aptly as the "republic of science."[2] To qualify for citizenship in the republic of science—to be accepted as a scientist—one's scientific credentials must be acceptable. Only those with proper credentials, as evidenced by past achievement in science, are allowed to participate in the government of science: only scientists are listened to in the mutual criticism that keeps science valid. What survives this criticism—or, as Harvey Brooks puts it, what has value in the "intellectual marketplace"—is incorporated in the corpus of science; all else is rejected.[3]

Citizens of the republic of science—that is, the scientists—are an elite within the larger society. Only scientists participate in the internal government of the republic of science, and the degree of participation of a scientist is determined by one's standing as a scientist. Where science and politics meet, however, issues can no longer be settled by scientists alone. The public, either directly or through articulate scientific pamphleteers who speak out as keepers of what they regard as the public interest, often engages in the debate. The issues affect everyone, not just scientists, and therefore everyone, in some sense, has a right to be heard. A biologist with no credentials in quantum electrodynamics would never think of attending a scientific meeting on that subject: not only would he be unable to understand it, his own scientific work would be untouched by it. By contrast, citizens of the most diverse scientific or educational qualification now participate in debate on repositories for radioactive wastes in salt mines, the dangers of pesticides, or the decision to build an SST. The obvious point is contained in the saying that he whose shoe pinches can tell something to the shoemaker.

The "republic of trans-science" (if one can identify something so diffuse as a republic) has elements of the political republic, on the one hand, and the republic of science on the other. Its character must therefore reflect to a great extent the political structure of the society in which it operates. In the United States, where the political tradition is strongly democratic and there is relatively little tradition of deference to authority, the debates on trans-scientific issues are particularly noisy. By contrast, in Western Europe, whatever debate occurs on such matters is more subdued, less open. In the Soviet Union, whatever debate occurs on these matters is practically inaudible in public.

What are the advantages and disadvantages of conducting the trans-scientific debate in a completely open manner, as is done in the United States? The disadvantages are clear, particularly to the experts. Often the line between scientific and trans-scientific issues is blurred: in fact, the essence of the matter is often to define just where the line between the two lies. Experts consider public intrusion into the scientific parts of the debate by the uninitiated as obfuscatory; on the other hand, the public's involvement helps force a delineation between science and trans-science.

To take an example, all would agree that, if every safety rod in a boiling-water reactor were to fail at the same time as the turbine tripped, a catastrophe would ensue, unless additional countermeasures were taken. This is a strictly scientific question that may be decided by the methods of science; and, in the case mentioned, the scientific facts are indisputable. To the experts, public discussion of this strictly scientific issue could only cause confusion, since science already gives an unequivocal answer.

On the other hand, the question of whether all the safety rods can ever fail simultaneously is trans-scientific. Here experts disagree, and the question is really unanswerable: although all who have studied the matter will agree that the probability of failure is extremely small, some will insist that the event is incredible, others that it is not. This second question, whether the postulated initiating event is credible, is trans-scientific: experts possessing sound credentials disagree. Here public discussion helps to remind us that science can say little about the matter and that its resolution requires nonscientific mechanisms.

The public discussion of trans-scientific questions, like the probability of a reactor accident, runs the risk of introducing exaggeration and distortion into the debate over reactor safety. Discussion of the simultaneous failure of all safety systems in a reactor at a crucially important juncture is essential to the technical assessment of the reactor—if for no other reason than to design countermeasures that will minimize the probability of such events ever taking place. Yet, taken out of context, such discussion can sometimes cause great confusion, if not panic. There develops an accumulation of contingency: each unlikely event connected with a reactor, once it becomes a matter of public discussion, seems to acquire a plausibility that goes much beyond what was originally intended when it was more cautiously formulated by scientifically trained persons. In consequence, reactors now, at least in the United States, are loaded down with safety system added to safety system—the safety and emergency systems almost dominate the whole technology.

By contrast, in the Soviet Union, where the public does not have an automatic right to be informed about, or to participate in, scientific and technological debate of this sort, the technology of reactors is less obviously centered on safety. Until recently, Soviet pressurized water reactors had no containment shells. Soviet engineers insisted that their primary systems were built so ruggedly that a catastrophic accident of the sort that the containment shell is intended to deal with was incredible; and, moreover, that the containment shell would not be effective if the accident were severe enough. There was here a divergence between the American and the Soviet views, both with respect to the effectiveness of containment shells and with respect to how safe is safe enough. One can attribute these differences simply to the existence of the very influential Advisory Committee on Reactor Safety in the United States. However, I would not underrate the importance of the difference in degree of access of the public to the technological debate in the Soviet Union and in the United States. In my view, the added emphasis on safety in the American systems is an advantage, not a disadvantage; and, insofar as this can be attributed to public participation in the debate over reactor safety, I would say such participation has been advantageous. Recently, the Soviet engineers have reconsidered the matter, and the newer Soviet pressurized water reactors are housed in containment shells. The extent to which the American debate has influenced this change in policy is hard for an outsider to judge.

What are the responsibilities of the scientist in trans-scientific debate? Although the scientist cannot provide definite answers to trans-scientific questions any more than can the lawyer, the politician, or a member of the lay public, the scientist does have one crucially important role: to make clear where science ends and trans-science begins. Now this is not at all easy, since experts will often disagree as to the extent and reliability of their expertise. Yet, as the current debates over the environment demonstrate, scientists often appear reluctant to concede limits to the proficiency of their science. The argument about low-level radiation insult would have been far more sensible had it been admitted at the outset that this was a question that went beyond science. The matter could then have been dealt with, initially, on moral or aesthetic grounds.

Beyond this, scientists possess a unique knowledge that borders the trans-scientific issues. It is this knowledge that they can and must use to inject discipline and order into the often chaotic trans-scientific debate. Even in trans-scientific debate, which inevitably weaves back and forth across the boundary between what is and what is not known and knowable, confrontation between scientists of opposing ethical or political positions is desirable. But, as the extraordinary debate over the antiballistic missile demonstrated,[4] scientists must exercise all the canons of scientific discipline in such a confrontation: if they do not adhere to these canons, then, as Robert L. Bartley has plaintively asserted, "If scientists do not scrupulously guard a certain minimum of detachment and self-restraint, what do they have to offer that the next man does not? If all questions are political, why not leave them all to the politicians?"[5]

The Impact of Trans-Science on the Republic of Science

Since the border between trans-science and science is elusive, it seems clear that the public will inevitably participate in debates that possess scientific as well as trans-scientific components. Could such participation in science by the uncertified tend to weaken the republic of science, citizenship in which is rigorously certified? If the public has a right to debate the details of reactor designs, then why not extend that right to the debate on whether nuclear physics or high-energy physics should be supported more heavily? If it has the right to debate the use of pesticides in agriculture, then should it not have the right to debate whether or not we should do experiments that might lead to human cloning or which might disclose racial differences in intelligence? And if the unaccredited public becomes involved in debates on matters as close to the boundary between science and trans-science as the direction of biological research, is there not some danger that the integrity of the republic of science will be eroded?

Another aspect of the public's intrusion in scientific debate is the tendency to expose such debate to public scrutiny, and thus to public debate, before "all the facts are in." A good example of this is the recent incident in the United States concerning the adequacy of emergency core-cooling systems in pressurized water reactors. The emergency core-cooling system springs into action in the very unlikely event that the regular cooling system fails to deliver water to the hot core of a reactor. Recent experiments at Idaho Falls raised questions about the reliability of the emergency system; but these experiments, performed on a very small scale, have been severely challenged by experts in the field.

The issue here is scientific, and it clearly can be answered by the usual mechanisms of science—experiment, additional analysis, challenge, and counterchallenge by those who have intimate knowledge of the matter. But because the emergency core-cooling system is so closely tied to the safety of reactors, the scientific debate has become a matter of intense public discussion and concern. Public pronouncements as to the outcome of the scientific argument are made, especially at hearings before reactor licensing boards; and the political pressure generated thereby may force an improper decision before the scientific debate has come to a proper conclusion.

Another possible danger of the public's involvement in scientific debate is illustrated by the Velikovsky incident.[6] To much of the public, I dare say, Velikovsky's treatment by the scientific community smacks of Galileo's treatment at the hands of the Inquisition. To a scientist, Velikovsky is not to be taken seriously because he did not conform to the rules of procedure of the republic of science; to the public, he is the victim of an arrogant elite. The nonscientific public came close, in the Velikovsky case, to demanding the right to pass judgement on scientific questions.

In the past, when science depended less completely upon the public for its support, it was perhaps not so serious that the public's views on scientific questions were ignored by the scientists. Today, however, one wonders whether science can afford the loss in public confidence that the Velikovsky incident cost it. The republic of science can be destroyed more surely by withdrawal of public support for science than by intrusion of the public into its workings.

That the republic of science may be compromised by encroachment from the public is probably an exaggeration: there will always be a part of science that is so unambiguously in the realm of science that it would be absurd to think of an encroachment by the uninitiated. But whether or not the republic of science is weakened, the die is cast. In the final analysis, no matter what the disadvantages of public access to technological and trans-scientific debate, I believe we have no choice. In a democratic society, the public's right of access to the debate, in the sense of being informed about it and participating in it, is as great as the public demands it to be. Especially where experts disagree, the public has little choice but to engage in the debate at an earlier stage than the experts themselves find convenient or comfortable. Questions with strong scientific content that impinge too forcefully on the public concern inevitably become incorporated in the republic of trans-science.

The late Harold Laski once said:

> . . . special knowledge and the highly trained mind produce their own limitations . . . . Expertise . . . sacrifices the insight of common sense to intensity of experience . . . . It has . . . a certain caste-spirit about it, so that experts tend to neglect all evidence which does not . . . belong to their own ranks . . . . where human problems are concerned, the expert fails to see that every judgment he makes not purely factual in nature brings with it a scheme of values which has no special validity about it.[7]

We in the technological and scientific community value our republic and its workings. But when what we do transcends science and when it impinges on the public, we have no choice but to welcome the public—even encourage the public—to participate in the debate. Scientists have no monopoly on trans-scientific wisdom: they will have to accommodate to the will of the public and its representatives. The republic of trans-science, bordering as it does on the political republic and the republic of science, can be neither as pure as the latter nor as undisciplined as the former. The most science can do is to inject some intellectual discipline into the republic of trans-science; politics in an open society will surely keep it democratic.

References

[1] H. P. Green, "The Risk Benefit Calculus in Nuclear Power Licensing", in *Nuclear Power and the Public,* edited by H. Foreman (University of Minnesota Press, Minneapolis, 1970), p. 131.

[2] M. Polanyi, Minerva **1** (1), pp. 54–73 (1962).

[3] H. Brooks, "The Federal Establishment for Science and Technology: Contribution to New National Goals," Conference on Research in the Service of Man: Biomedical Knowledge, Development, and Use, sponsored by the U.S. Senate Committee on Government Operations, Oklahoma City (U.S. Government Printing Office, Washington, DC, 1967), pp. 57–64; *The Government of Science* (The MIT Press, Cambridge, MA, 1968).

[4] "The Obligations of Scientists as Counsellors: Guidelines for the Practice of Operations Research," Minerva **10** (1), pp. 107–157 (1972).

[5] R. L. Bartley, "When Science Tangles with Politics," *The Wall Street Journal* (October 12, 1971).

[6] Immanuel Velikovsky created a sensation during the 1950s in the United States with his book, *Worlds in Collision* (Doubleday, Garden City, NY, 1950); M. Polanyi, Minerva **5** (4), pp. 533–545 (1967).

[7] "The Limitations of the Expert," *Harper's Magazine* (December, 1930), *op. cit.*, p. 136.

The Regulator's Dilemma

In his essay, "Risk, Science, and Democracy," William D. Ruckelshaus expresses very clearly what I call the regulator's dilemma. During the past 15 years, Ruckelshaus notes, there has been a shift in public emphasis from visible and demonstrable pollution problems, such as smog resulting from automobiles and the handling of raw sewage, to potential and largely invisible problems, such as the effects of low concentrations of toxic pollutants on human health. This shift is important for two reasons. First, it has changed the way science is applied to practical questions of public health protection and environmental regulation. Second, it has raised difficult questions about managing chronic risks within the context of free and democratic institutions.[1]

When the environmental concern was patent and obvious—such as the problem of smog in Los Angeles—science could and did provide unequivocal answers. Smog, for example, comes from the gas emissions from burning liquid hydrocarbons, and the answer to the smog problem lies in controlling those emissions. The regulator's course was rather straightforward because the science upon which regulatory decisions are made was operating well within its power. However, when the environmental concern is subtle—for example, how much cancer is caused by an increase of 10 percent in mean background radiation—science is being asked a question that lies beyond its power; the question is trans-scientific. Yet the regulator, by law, is expected to regulate

Adapted from a paper at a National Academy of Engineering Symposium on "Hazards, Technology and Fairness," 1985. First published in Issues in Science and Technology **2**(1), pp. 59–72 (1985). Reprinted by permission.

even though science can hardly help him; this is the regulator's dilemma.

Many issues arise in the adjudication of disputes over who is to blame and who is to be compensated for damage allegedly caused by rare events, such as nuclear accidents. The regulator's dilemma is also faced by the judge who is presiding over a tort case involving, for example, a claim for damages blamed on a toxic waste dump. Indeed, the regulator's dilemma could equally be called the toxic tort dilemma.

A lawsuit alleging injury from chemical pollutants is unlike the traditional liability case. If my car injures a pedestrian, I am liable to be sued. What is at issue, however, is not whether I have injured a pedestrian. Rather, it is whether I am at fault. On the other hand, if the lead from my car's exhaust is alleged to cause bodily harm, the issue is not whether my car emitted the lead but whether the lead actually caused the alleged harm. The two situations are quite different. In the first example, the relation between cause and injury is not at issue. In the second, it is the issue.

II

Science deals with regularities in our experience; art deals with singularities. It is no wonder that science tends to lose its predictive or even its explanatory power when the phenomena it deals with are singular, irreproducible, and one of a kind—in other words, rare. Although science can often analyze a rare event after the fact—for example, the extinction of dinosaurs at the end of the Cretaceous period following the presumed collision of the Earth and an asteroid—it has great difficulty predicting when such an uncommon event will occur.

I distinguish here between two sorts of rare events—accidents and low-level insults, whose potential to cause injury is unknown. Accidents are large-scale malfunctions whose etiology is not in doubt but whose likelihood is very small. The partial nuclear reactor meltdown at Three Mile Island in 1979 and the release of toxic gas from a chemical plant at Bhopal, India, in 1984 are examples of accidents. The precursors to these specific events—for example, the condition of the auxiliary feedwater system and other components at Three Mile Island—and the way in which the accidents unfolded are well understood. Estimates of the likelihood of the particular sequence of malfunctions are less firmly grounded. As the number of individual accidents increases, prediction of their probability becomes more and more reliable. We can predict very well how many automobile fatalities will occur in 1986; we can hardly claim the same degree of reliability in predicting the number of serious reactor accidents in 1986.

Low-level insults are rare in a rather different sense. We know that about 100 rems of radiation will double the mutation rate in a large population of exposed mice. How many mutations will occur in a population of mice exposed to 100 millirems of radiation? In this case, the mutations—if induced at all by such low levels of exposure—are so rare that to demonstrate an effect with 95 percent confidence would require the examination of many millions of mice. Although such an effort is not impossible in principle, it is in practice. Moreover, even if we could perform so heroic a mouse experiment, the extrapolation of these findings to humans would still be fraught with uncertainty. Thus, human injury from low-level exposure to radiation is a rare event whose frequency cannot be accurately predicted.

III

When dealing with events of this sort, science resorts to the language of probability. Instead of saying that this accident will happen on that date, or that a particular person exposed to a low-level dose of radiation will suffer a particular fate, it tries to assign probabilities for such occurrences. Of course, where the number of instances is very large or the underlying mechanisms are fully understood, the probabilities are themselves perfectly reliable. In quantum mechanics, there is no uncertainty about the probability distribution of the phenomenon being described. In the class of phenomena considered here, however, even though the likelihood of an event happening or of a disease being caused by a specific exposure is given as a probability, the probability itself is very uncertain. One can think of a somewhat fuzzy demarcation between what I have called science and trans-science. The domain of science covers phenomena that are deterministic or whose probability of occurrence can itself be stated precisely; in contrast, trans-science covers those events whose probability of occurrence is itself highly uncertain.

Despite the difficulties, scientific mechanisms have been devised for estimating, however imperfectly, the probability of rare events. For accidents, the technique is probabilistic risk assessment (PRA); for low-level insults, various empirical and theoretical approaches are used.

Although probabilistic risk assessment had been used in the aerospace industry for a long time (for example, to predict the reliability of components), it first sprang into public prominence in 1975 with a reactor-safety study directed by nuclear engineer Norman C. Rasmussen.[2] The Rasmussen study, sponsored by the Atomic Energy Commission (and then by the Nuclear Regulatory Commission), was designed to estimate the public risks involved in potential accidents at commercial nuclear reactors.

Probabilistic risk assessment, when applied to nuclear reactors, seeks to identify all sequences of subsystem failures that may lead to a failure of the overall system; it then tries to estimate the consequences of each subsystem failure so identified. The result is a probability distribution, $P(C)$; that is, the probability, P, per reactor year, of a consequence having magnitude C. Consequences include both material damage and health effects. Usually, the probability of accidents having large consequences is less than the probability of accidents having small consequences.

A probabilistic risk assessment for a reactor requires two separate estimates: first, an estimate of the probability of each accident sequence; second, an estimate of the consequences—particularly the damage to human health—caused by the uncontrolled radioactive effluents released in the accident. An accident sequence is a series of equipment or human malfunctions, such as a pump that fails to start, a valve that does not close, or an operator confusing an ON with an OFF signal. We have statistical data for many of these individual events; for example, enough valves have operated for enough years so that we can, at least in principle, make pretty good estimates of the probability of failure.

Uncertainties still remain, however, because we can never be certain that we have identified every relevant sequence. Proof of the adequacy of probabilistic risk assessment must therefore await the accumulation of operating experience. For example, the median probability of a core melt in a light-water reactor, according to the 1975 Rasmussen study, was 1 in every 20,000 reactor-years; the core melt at Three Mile Island's number two reactor (TMI-2) occurred after only 700 reactor-years. The number two reactor, however, differed from the reactors Rasmussen studied, and, in retrospect, one could rationalize most of the discrepancy between his estimate and the seemingly premature occurrence at TMI-2.

Since the core melt at Three Mile Island, the world's light-water reactors by 1991 have accumulated some 3,000 reactor-years of operation without a core melt. This performance places an upper limit on the *a priori* estimate of the core-melt probability. Thus, if this probability were as high as 1 in every 1,000 reactor years, the likelihood of surviving 3,000 reactor-years would not be more than 5 percent; put otherwise, we can say with 95 percent confidence that the core-melt probability is not as high as 1 in 1,000 reactor years. With about 400 light-water reactors on line in the world, we would accumulate about 7,000 reactor-years by the year 2000. Should we survive without a core melt, we could then say, with 95 percent confidence, that the core-melt probability is not higher than 1 in 2,300 reactor-years. In the absence of such experience, one is left with rather subjective judgments. Although Harold W. Lewis, in his critique of Rasmussen's 1975 study,[3] asserts that he could not place a bound on the uncertainty of probabilistic risk assessment, Rasmussen argued that his estimate of core-melt probability may be in error by about a factor of 10 either way—that is, the probability may be as high as 1 in 2,000 reactor-years or as low as 1 in 200,000 reactor-years.

Thus, after 7,000 reactor-years of operation without a core melt, we can say, with better than 95 percent confidence, that Rasmussen's upper limit (1 in 2,000 reactor-years) is not too optimistic. Our confidence in probabilistic risk assessment can eventually be tested against actual, observable experience. Until this experience has been accumulated, however, we must concede that any probability we predict must be highly uncertain. To this degree, our science is incapable of dealing with rare accidents, but time, so to speak, annihilates uncertainty in estimates of accident probability.

Unfortunately, time does not annihilate uncertainties over consequences as unequivocally as it does uncertainties over frequency of accidents. A large reactor or chemical-plant accident can cause both immediate, acute health effects and delayed, chronic health effects. If the exposure either to radiation or to methyl isocyanate is high enough, the effect on health is quite certain. For example, a single exposure of about 400 rems will cause about half of the people exposed to die. On the other hand, in a large accident, many people will also be exposed to smaller doses—indeed, to doses so low that the resulting health effects are undetectable. At Bhopal, many thousands of people were exposed to methyl isocyanate but most of them recovered. We cannot say positively whether or not they will suffer some chronic disability.

The very worst accident envisaged in the Rasmussen study, with a probability of 1 in a billion reactor-years, projected an estimated 3,300 early fatalities, 45,000 early illnesses, and 1,500 delayed cancers per year among 10 million exposed people. Almost all of the estimated delayed cancers are attributed to exposures of less than 1,000 millirems per year—a level at which we are very hard put to estimate the risk of inducing cancer. Similarly, the American Physical Society's critique of the Rasmussen study attributed an additional 10,000 deaths over 30 years among 10 million people exposed to cesium-135 dispersed in a very large accident.[4] The average exposure in this case was assumed to be 250 millirems per year—again, a level at which our estimates of the health effects are extremely uncertain.

Has the nuclear community, particularly its regulators, figuratively shot itself in the foot by trying to estimate the number of delayed casualties as a result of these low-level exposures? In retrospect, I think the Rasmussen study would have been on more solid ground had it confined its estimates to those health effects resulting from exposures at higher levels, where science makes reliable estimates. For the lower exposures, the consequences could have been stated simply as the number of man-rems (the number of people multiplied by the number of rems) of exposure of individuals whose total exposure did not exceed, say, 5,000 millirems, without trying to convert this man-rems number into numbers of latent cancers. Thus, health consequences would be reported in two categories: (1) for highly exposed individuals, the number of health effects; and (2) for slightly exposed individuals, the total man-rems, or even the distribution of exposures accrued by the large number of individuals so exposed. Perhaps such a scheme could be adopted in reporting the results of future probabilistic risk assessments; at least it has the virtue of being more faithful than is the present convention to the state of scientific knowledge.

IV

In both of my examples of accidents (Bhopal and nuclear accidents), many people are exposed to low-level insult. The uncertainties inherent in estimating the effects of such low-level exposure are heaped on top of the uncertainties in estimating the probability of the accident that may lead to exposure in the first place.

Science has exerted great effort to ascertain the shape of the dose-response curve at low dose—but very little, if anything, can be said with certainty about the low-dose response. To quote the report of the National Research Council, *The Effects on Populations of Exposure to Low Levels of Ionizing Radiation: 1980* (also known as BEIR-III, for the committee that prepared it, the Committee on the Biological Effects of Ionizing Radiation), "The Committee does not know whether dose rates of gamma or x rays of about 100 mrads/yr are detrimental to man . . . . It is unlikely that carcinogenic and teratogenic effects of doses of low-LET radiation administered at this dose rate will be demonstrable in the foreseeable future."[5] This prompted Philip Handler, then president of the National Academy of Sciences, to comment in his letter of transmittal to the Environmental Protection Agency, which had requested the study, "It is not unusual for scientists to disagree . . . [and] the sparser and less reliable the data base, the more opportunity for disagreement . . . . The report has been delayed . . . to permit time . . . to display all of the valid opinions rather than distribute a report that might create the false impression of a clear consensus where none exists."[6]

This forthright admission—that science can say little about low-level insults—I find admirable. It represents an improvement over the unjustified assertion in the BEIR-I report of 1972 that 170 millirems per year over 30 years, if imposed on the entire U.S. population, would cause between 3,000 and 15,000 cancer deaths per year.[7] I do not quarrel with the estimated upper limit—which amounts to one cancer per 2,500 man-rems, but I think placing the lower limit at 3,000 rather than at zero is unjustified. Moreover, I think it has caused great harm. The proper statement should have been that, at 170 millirems per year, we estimate the upper limit for the number of cancers to be 15,000 per year; the lower limit may be zero.

Since the appearance of the BEIR reports, two other developments have added to the burden of those who must judge the carcinogenic hazard of low-level insults: an awareness and study of (1) natural carcinogens and (2) ambiguous carcinogens.

Natural carcinogens

Is cancer environmental, in the sense of being caused by technology's effluents, or is it a natural consequence of aging? In the past few years, we have seen a remarkable shift in viewpoint: 15 years ago, most cancer experts would have accepted a primarily environmental etiology for cancer, but today the view that natural carcinogens are far more important than are man-made ones has gained many converts. In his 1983 article in *Science*, biochemist Bruce N. Ames marshaled powerful evidence that many of our most common foods contain naturally occurring carcinogens.[8] Indeed, biochemist John R. Totter, former director of the Atomic Energy Commission's division of biology and medicine, has offered evidence for the oxygen-radical theory of carcinogenesis: that we eventually get cancer because we metabolize oxygen and subsequently produce oxygen radicals that can play havoc with our DNA.[9] As such views of the etiology of cancer acquire scientific support, I think that the trans-scientific question about how much cancer is caused by a tiny chemical or physical insult will be recognized as irrelevant. One does not swat gnats when pursued by elephants.

Ambiguous carcinogens

To further complicate the cancer picture, there is evidence that some agents, such as dioxin, various dyes, and even moderate levels of radiation, seem to diminish the incidence of some cancers while simultaneously increasing the incidence of others. The lifespan of the animals exposed to those agents in laboratory tests on average exceeds that of animals not exposed.[10] A most striking example, given by biostatistician Joseph K. Haseman, is yellow dye number 14 given to leukemia-prone female rats. This dye completely suppresses leukemia, which is always fatal, but causes liver tumors, most of which are benign.

I mention these two findings—or perhaps they should be considered points of view—to stress my underlying point: when we are concerned with low-level insult to human beings, we can say very little about the cancer dose-response curve. Saying that so many cancers will be caused by so much low-level exposure to so many people, a practice that terrifies many people, goes far beyond what science actually can say.

V

Does the scientific community accept the notion that there are intrinsic limits to what it can say about rare events—that, as events become rarer, the uncertainty in the probability of occurrence of a rare event is bound to grow? Perhaps a better way of framing this question is: To what use can we put scientific tools of investigation of rare events, such as probabilistic risk assessment and large-scale animal experiments, if we concede that we can never get definitive answers?

I believe that probabilistic risk assessment with an uncertainty factor as high as 10 is often useful, especially if one uses the technique for comparing risks. For example, the 3,000 reactor-years already experienced since the Three Mile Island accident suggest that the probability of a core melt is likely to be less than 1 in 1,000 reactor-years and may well be as low as less than 1 in 10,000 reactor-years. This is to be compared with dam failures whose probability, based on many hundreds of thousands of dam-years (where time has annihilated uncertainty), is around 1 in 10,000 dam-years. Even with an uncertainty factor of 10, we can judge how safe reactors are compared to dams.

When one compares the relative intrinsic safety of two very similar devices—such as two water-moderated reactors—probabilistic risk assessment is on much more solid ground. Here one is asking not for absolute estimates of risk but rather for estimates of relative safety. If reactors A and B differ in only a few details—say reactor A has two auxiliary feedwater trains whereas B has only one—the ratio of core-melt probabilities should be much more reliable than their absolute values because the ratio requires an estimate of failure of a single subsystem—in this instance, the extra auxiliary feedwater on reactor A.

Not only can one say with reasonable assurance how much safer reactor A is than reactor B, but, as a result of the detailed analysis, one can identify the subsystems that contribute most to the estimated failure rate. Even if probabilistic risk assessment is inaccurate, it is very useful in unearthing deficiencies; one can hardly deny that a reactor in which deficiencies revealed by probabilistic risk assessment have been corrected is safer than one in which they have not been corrected, even if one is unwilling to say how much safer.

Somewhat the same considerations apply to low-level insult. An agent that does not shorten lifespan at high dose will not shorten lifespan at low dose. An agent that is a very powerful carcinogen at high dose is more likely to be a carcinogen at low dose than one that is a less powerful high-dose carcinogen. (This belief, however, has recently been challenged by Bruce Ames. At high doses, an agent is mitogenic, even if it is *not* mutogenic; and mitogenesis—i.e., replacement of killed cells—is carcinogenic.) Animal experiments are useful in deciding which agents to worry about and which not to worry about. Of course, the Ames test (which determines by a relatively simple procedure whether a substance is mutagenic) has made at least some preliminary screening of carcinogens more feasible because substances that cause mutations are considered to be potential carcinogens. The difficulty today seems to be not so much identifying agents that at high dose may be carcinogens as it is prohibiting exposures far below levels at which no effect can be, or perhaps ever will be, demonstrated. The regulator and the concerned citizen are inclined to approve the Delaney clause of the Federal Food, Drug, and Cosmetic

Act, which prohibits the use of any food additive that has been shown to cause cancer in laboratory animals or humans. This clause, however, is of no help in resolving such issues as the relative risks of, say, cancer induction by nitrosamines (carcinogenic compounds that can be formed in the body from nitrites) and digestive disorders caused by meat untreated with nitrites.

The Delaney clause is the worst example of how a disregard of an intrinsic limit of science can lead to bad policy by overenthusiastic politicians. Harvard physicist Harvey Brooks has often pointed out that one can never prove the impossibility of an event that is not forbidden by a law of nature. Most will agree that a perpetual motion machine is impossible because it violates the laws of thermodynamics. That one molecule of a polychlorinated biphenyl (PCB) may cause a cancer in humans is a proposition that violates no law of nature: hence many, even within the scientific community, seem willing to believe that this possibility is something to worry about. It was this error that led to the Delaney clause.

When is an event so rare that the prediction of its occurrence forever lies outside the domain of science and therefore within the domain of trans-science? Clearly, we cannot say; and perhaps, as science progresses, this boundary between science and trans-science will recede toward events of lower frequency. At any stage, however, the boundary is fuzzy, and much scientific controversy rages over deciding where it lies. One need only read the violent exchange between Edward P. Radford and Harald H. Rossi over the risk of cancer from low levels of radiation to recognize that, where the facts are obscure, argument—even *ad hominem* argument—blossoms.[11] Indeed, Alice Whittemore, in her "Facts and Values in Risk Analysis for Environmental Toxicants," has pointed out that facts and values are always intermingled at this "rare event" boundary between science and trans-science.[12] A scientist who believes that nuclear energy is evil because it inevitably leads to proliferation of nuclear weapons (which is a common basis for opposition to nuclear energy) is likely to judge the data on induction of leukemia from low-level exposures at Nagasaki differently than is a scientist whose whole career has been devoted to making nuclear power work. Cognitive dissonance is all but unavoidable when the data are ambiguous and the social and political stakes are high.

VI

No one would dispute that judgments of scientific truth are much affected by the scientist's value system when the issues are at or close to the boundary between science and trans-science. On the other hand, as the matter under dispute approaches the domain of science, most would claim that the scientist's extrascientific values intrude less and less. Soviet scientists and U.S. scientists may disagree on the effectiveness of a ballistic-missile defense, but they agree on the cross section of U-235 or the lifetime of the pi meson.

This all seems obvious, even trite. Yet, in the past decade or so, a school of sociology of knowledge has sprung up in Great Britain that claims that "scientific views are determined by social (external) conditions, rather than by the internal logic of scientific tradition and inherent characteristics of the phenomenal world,"[13] or that "all knowledge and knowledge claims are to be treated as being socially constructed: genesis, acceptance, and rejection of knowledge [is] sought in the domain of the Social World rather than . . . the Natural World."[14]

The attack here is not on science at the boundary with trans-science—in particular, the prediction of the frequency of rare events. At least the more extreme of the sociologists of knowledge claim that using traditional ways of establishing scientific truth—by appealing to nature in a disciplined manner—is not how science really works. Scientists are seen as competitors for prestige, pay, and power, and it is the interplay among these conflicting aspirations, not the working of some underlying scientific ethic, that defines scientific truth. To be sure, these attitudes toward science are not widely held by practicing scientists; however, they are taken seriously by many political activists who, though not in the mainstream of science, nevertheless exert important influence on other institutions—the press, the media, the courts—that ultimately influence public attitudes toward science and its technologies.

If one takes such a caricature of science seriously, how can one trust a scientific expert? If scientific truth, even at the core of science, is decided by negotiation between individuals in conflict because they hold different nonscientific beliefs, how can one say that this scientist's opinion is preferable to that one's? Furthermore, if the matter at issue moves across the boundary between science and trans-science, where all we can say with certainty is that uncertainties are very large, how much less able are we to distinguish between the expert and the charlatan, between the scientist who tries to adhere to the usual norms of scientific behavior and the scientist who suppresses facts that conflict with his political, social, or moral preconceptions?

One way to deal with these assaults on scientists and scientific truth would be to define a new branch of science, called regulatory science, in which the norms of scientific proof are less demanding than are the norms in ordinary science. I should think that a far more honest and straightforward way of dealing with the intrinsic inability of science to predict the occurrence of rare events is to concede this limitation and not to ask of science or scientists more than they are capable of providing. Instead of asking science for answers to unanswerable questions, regulators should be content with less far-reaching answers. For example, where the ranges of uncertainty can be established, regulate on the basis of uncertainty; where the ranges of uncertainty are so wide as to be meaningless, recast the question so that regulation does not depend on answers to the unanswerable. Furthermore, because these same limits apply to litigation, the legal system should recognize, much more explicitly than it has, that science and scientists often have little to say, probably much less than some scientific activists would admit.

The expertise of scientific adversaries is often at the heart of litigation over personal injury alleged to be caused by subtle, low-level exposures. Each side presents witnesses whose scientific credentials it regards as impeccable. Because the issues themselves tend to be trans-scientific, one can hardly decide the validity of the assertions of either side's witnesses. Under the circumstances, I suppose, one is justified in regarding scientific witnesses no differently than other witnesses; their credibility is judged by past records, behavior, and general demeanor, as well as the self-consistency of their testimony. Such, at least, was the way in which a federal district court judge, Patrick Kelley, settled *Johnston v. United States*, in which the issue was the claim that exposure to radiation from reworking old aircraft instrument dials had caused injury; Kelley impugned, on grounds no different from those one would invoke in an ordinary lawsuit, the competence if not the integrity of some of the plaintiff's scientific witnesses.

VII

There are various ways to provide some assurance of safety despite uncertainty. Here I briefly describe two of these ways—which I call the technological fix and *de minimis*—without claiming that these are the most important ones, let alone the only ones.

Technological fix

Science cannot exactly predict the probability of a serious accident in a light-water reactor or the likelihood that a radioactive-waste canister in a depository will dissolve and release radioactivity to the environment. Can one design reactors or waste canisters for which the probability of such occurrences is zero—or at least, where the prevention of such mishaps relies on immutable laws of nature that can never fail rather than on the less than reliable intervention of electromechanical devices? Surprisingly, this approach to nuclear safety has come into prominence only in the past five years. Kåre Hannerz in Sweden and Herbert Reutler and Günter H. Lohnert in West Germany have proposed reactor systems whose safety depends not on active interventions but rather on passive, inherent characteristics.[15] Although one cannot say that the probability of mischance has been reduced to zero, there is little doubt that the probabilities are several, perhaps three, orders of magnitude lower than the probabilities of mischance for existing reactors. To the extent that such proposed reactors embody the principle of inherent safety, their adoption would avoid much of the dispute over reactor safety, the limits on nuclear accident liability contained in the Price-Anderson Act, repetition of the Three Mile Island accident, and so forth. In short, such a technological fix enables one largely to ignore the uncertainties in any prediction of core-melt probabilities.

The idea of incorporating inherent or passive safety into the design of chemical plants had been proposed by Trevor A. Kletz of the Loughborough University of Technology in 1974, shortly after the disaster at the Flixborough cyclohexane plant, which killed 28 people.[16] I suspect that one of the main consequences of the Bhopal disaster will be the incorporation of inherent safety features into new chemical plants; again, a way of finessing uncertainty in predicting failure probabilities.

De minimis

A perfect technological fix, such as a totally safe reactor or a crash-proof car, is usually not available, at least not at an affordable cost. Some low-level exposure to materials that are toxic at high levels is inevitable, even though we can never accurately establish the risk of such exposure. One way of dealing with this situation is to invoke the principle of *de minimis*. This principle, as Howard I. Adler and I suggested several years ago, argues that, for insults that occur naturally and to which the biosphere has always been exposed and presumably to which it has adapted, one should not worry about any additional manmade exposure as long as the manmade exposure is small compared to the natural exposure.[17] The basic idea is that the natural level of a ubiquitous exposure (such as cosmic radiation), if it is deleterious, cannot have been very deleterious because, in spite of its ubiquity, humans have survived it. Moreover, we do not know—and can never know—what the residual effect of that natural exposure really is. An additional exposure that is small compared to natural background radiation should be acceptable; at the very least, its deleterious effect, if any, cannot be determined.

Adler and I suggested that, for radiation whose natural background is well known, one may choose a *de minimis* level as the standard deviation of the natural background. This turns out to be around 20 percent of the mean background, around 20 millirems per year; this value has been used as the Environmental Protection Agency standard for exposure to the entire radiochemical fuel cycle.

Scientists know more about the natural incidence and biological effects of radiation than they do about any other agent. It would be natural, therefore, to use the standard established for radiation as a standard for other agents. This approach has been used by chemist T. Westermark of the Royal Institute of Technology in Stockholm. He has suggested that, for such naturally occurring carcinogens as arsenic, chromium, and beryllium, one may choose a *de minimis* level to be, say, 10 percent of the natural background.[18]

Clearly, a *de minimis* level will always be somewhat arbitrary. Nevertheless, it seems to me that, unless such a level is established, we shall forever be involved in fruitless arguments, the only beneficiaries of which will be the toxic-tort lawyers. Could the principle of *de minimis* be applied in litigation in much the same way it may be applied to regulation—that is, if the exposure is below *de minimis*, then the blame is intrinsically unprovable and cannot be litigated? I would imagine that the legal *de minimis* may be set higher than the regulatory *de minimis*; for example, the legal *de minimis* for radiation could be the background (after all, the BEIR-III committee concedes there is no way of knowing whether or not such levels are deleterious). The regulatory *de minimis* could justifiably be lower, simply on grounds of erring on the side of safety.

One approach may be to concede that there is some level of exposure that is beyond demonstrable effect. This defines a trans-scientific threshold. A *de minimis* level could then be established at some fraction—say, one-tenth—of this beyond-demonstrable-effect level. For example, if we take 100 millirems per year of radiation as the beyond-demonstrable-effect level for general somatic effects (damaging somatic cells as opposed to germ-line cells), which is the value according to the BEIR-III committee, a *de minimis* level could be set at 10 millirems per year. Of course, such a procedure would evoke much controversy as to what is the beyond-demonstrable-effect level or whether 10 is an ample safety factor. This example demonstrates, however, that, at least in the case of low-level radiation, a scientific committee has been able to agree on a beyond-demonstrable-effect level. As for the safety factor of 10, this cannot be adjudicated on scientific grounds. The most one can say is that tradition often supports a safety factor of 10—for example, the old standard for public exposure (500 millirems per year) was set at one-tenth of the tolerance level for workers (5,000 millirems per year).

Can the principle of *de minimis* be applied to accidents? What I have in mind is the notion that accidents that are sufficiently rare may be regarded somehow in the same category as acts of God and be compensated accordingly. We already recognize that natural disasters should be compensated by the society as a whole. One can argue that an accident whose occurrence requires an exceedingly unlikely sequence of untoward events may also be regarded as an act of God. Thus, the Price-Anderson Act could be modified so that, quite explicitly, accidents whose consequences exceeded a certain level, and whose probability as estimated by probabilistic risk assessment would be less than, say, 1 in 1 billion per year, would be treated as acts of God. Compensation in excess of the amount stipulated in the revised act would be the responsibility of Congress. The cutoff either for compensation or for probabilities would be negotiable, and perhaps it would be revised every 10 years or so. One not entirely fanciful suggestion may be to set any probability of the order of 1 in 10 million to 1 in 100 million per year to be a *de minimis* cutoff, this being the frequency at which the Earth may have been visited by the cometary asteroids that may have caused the extinction of species in past geologic eras.

VIII

As in most such questions, identifying and characterizing the problem is easier than solving it. That the dilemma of the regulator and the toxic-tort judge is rooted in science's inability to predict rare events cannot be denied. Getting the regulator and the toxic-tort judge off the horns of the dilemma is far from easy, and my two suggestions—the technological fix and *de minimis*—are offered tentatively and with diffidence.

Equally obvious is the intrinsic social dimension of the issue. In an open, litigious democracy such as ours, any regulation and any judicial decision can be appealed, and if the courts offer no redress, Congress, in principle, can do so. These legal mechanisms are ponderous, however. The result seems to me to be a gradual slowing of our technological-social engine as we become more and more enmeshed in fruitless argument over unresolvable questions.

Western society was debilitated once before by such fruitless, quixotic tilting with windmills. I refer of course to the devastating campaign against witches from the fourteenth century to the seventeenth century. As ecologist William Clark has put it so vividly, society took it for granted during that period that death, disease, and crop failure could be caused by witches.[19] To avoid such catastrophes, one had to burn the witches responsible for them—and consequently some million innocent people were burned. Finally, in 1610, the Spanish inquisitor Alonzo Salazar y Frias realized there was no demonstrated connection between catastrophe and witches. Although he did not prohibit the burning of witches, he did prohibit use of torture to extract confessions. The burning of witches, and witch hunting generally, declined precipitously.

I have recounted this story many times by now. Yet it still seems to me to capture the essence of our dilemma: the connection between low-level insult and bodily harm is probably as difficult to prove as the connection between witches and failed crops. I regard it as an aberration that our society has allowed this issue to emerge as a serious social concern, which in the modern context is hardly less fatuous than were the witch hunts of the past. That dark phase in western society died out only after several centuries. I hope our open, democratic society can regain its sense of proportion far sooner and can get on with managing the many real problems we always will face rather than waste its energies on essentially insoluble—and, by comparison, intrinsically unimportant—problems.

References

[1] W. D. Ruckelshaus, "Risk, Science, and Democracy," *Issues in Science and Technology* **1**, 19–38 (1985).

[2] U.S. Nuclear Regulatory Commission, *Reactor Safety Study: An Assessment of Accident Risk in U.S. Commercial Nuclear Plants* (WASH-1400, NUREG 75/014) (Washington, DC, 1975).

[3] U.S. Nuclear Regulatory Commission, *Risk Assessment Review Group Report to the U.S. Nuclear Regulatory Commission* (NUREG/CR-0400) (Washington, DC, September 1978), p. vi.

[4] "Report to the American Physical Society by the Study Group on Light Water Reactor Safety," Rev. Mod. Phys. **47** (Supplement 1) (1975).

[5] National Research Council, *The Effects on Populations of Exposure to Low Levels of Ionizing Radiation: 1980* (BEIR-III) (National Academy Press, Washington, DC, 1980), p. 3.

[6] National Research Council, *ibid.*, p. iii.

[7] National Research Council, *The Effects on Populations of Exposure to Low Levels of Ionizing Radiation* (BEIR-I) (National Academy Press, Washington, DC, 1972), p. 2.

[8] B. N. Ames, "Dietary Carcinogens and Anticarcinogens," Science **221**, 1249, 1256–64 (1983).

[9] J. R. Totter, "Spontaneous Cancer and Its Possible Relationship to Oxygen Metabolism," *Proceedings of the National Academy of Sciences* **77**, 1763–67 (1980).

[10] A. M. Weinberg and J. B. Storer, "On 'Ambiguous' Carcinogens and Their Regulation," Risk Analysis **5**, 151–155 (1985).

[11] National Research Council, *The Effects on Populations of Exposure to Low Levels of Ionizing Radiation: 1980* (BEIR-III), 287-321.

[12] A. Whittemore, "Facts and Values in Risk Analysis for Environmental Toxicants," Risk Analysis **3**, 23–33 (1983).

[13] J. Ben-David, Sociological Inquiry **48** (3,4), 209 (1978).

[14] T. J. Pinch and W. E. Bijker, Social Studies of Science **14**, 401 (1984).

[15] K. Hannerz, *Towards Intrinsically Safe Light Water Reactors* [ORAU/IEA-83-2(M) Rev.] (Oak Ridge, Tenn.: Oak Ridge Associated Universities, Institute for Energy Analysis, June 1983); H. Reutler and G. H. Lohnert, Nuclear Technology **62**, 22–30 (1983).

[16] T. A. Kletz, *Cheaper, Safer Plants or Wealth and Safety at Work, Notes on Inherently Safer and Simpler Plants* (The Institution of Chemical Engineers, Rugby, England, 1984).

[17] H. I. Adler and A. M. Weinberg, Health Physics **34**, 719–20 (1978).

[18] T. Westermark, *Persistent Genotoxic Wastes: An Attempt at a Risk Assessment* (Royal Institute of Technology, Stockholm, 1980).

[19] W. C. Clark, *Witches, Floods, and Wonder Drugs, Historical Perspectives on Risk Management* (RR-81-3) (International Institute for Applied Systems Analysis, Laxenburg, Austria, 1981).

Can Technology Replace Social Engineering?

During World War II and immediately afterward, our federal government mobilized its scientific and technical resources, such as the Oak Ridge National Laboratory, around great technological problems. Nuclear reactors, nuclear weapons, radar, and space are some of the miraculous new technologies that have been created by this mobilization of federal effort.

In the past few years, there has been a major change in focus of much of our federal research. Instead of being preoccupied with technology, our government is now mobilizing around problems that are largely social. We are beginning to ask what can we do about world population, about the deterioration of our environment, about our educational system, our decaying cities, race relations, poverty. President Johnson has dedicated the power of a scientifically oriented federal apparatus to finding solutions for these complex social problems.

Social problems are much more complex than are technological problems. It is much harder to identify a social problem than a technological problem: how do we know when our cities need renewing, or when our population is too big, or when our modes of transportation have broken down? The problems are, in a way, harder to identify just because their solutions are never clear-cut: how do we know when our cities are renewed, or our air clean enough, or our transportation convenient enough? By contrast, the availability of a crisp and beau-

Acceptance speech for the University of Chicago Alumni Medal, June 11, 1966. First published in the University of Chicago Magazine, Volume 59, No. 1, October, 1966, pp. 6–10. Reprinted by permission.

tiful technological solution often helps focus on the problem to which the new technology is the solution. I doubt that we would have been nearly as concerned with an eventual shortage of energy as we now are if we had not had a neat solution—nuclear energy—available to eliminate the shortage.

There is a more basic sense in which social problems are much more difficult than are technological problems. A social problem exists because many people behave, individually, in a socially unacceptable way. To solve a social problem, one must induce social change—one must persuade many people to behave differently than they have behaved in the past. One must persuade many people to have fewer babies, or to drive more carefully, or to refrain from disliking Negroes. By contrast, resolution of a technological problem involves many fewer individual decisions. Once President Roosevelt decided to go after atomic weapons, it was a relatively simple task to mobilize the Manhattan Project.

The resolution of social problems by the traditional methods—by motivating or forcing people to behave more rationally—is a frustrating business. People don't behave rationally; it is a long, hard business to persuade individuals to forego immediate personal gain or pleasure (as seen by the individual) in favor of longer-term social gain. And, indeed, the aim of social engineering is to invent the social devices—usually legal, but also moral, educational, and organizational—that will change each person's motivation and redirect his activities in ways that are more acceptable to society.

The technologist is appalled by the difficulties faced by the social engineer; to engineer even a small social change by inducing individuals to behave differently is always hard, even when the change is rather neutral or even beneficial. For example, some rice eaters in India are reported to prefer starvation to eating wheat that we send to them. How much harder it is to change motivations where the individual is insecure and feels threatened if he acts differently, as illustrated by the poor white's reluctance to accept the Negro as an equal. By contrast, technological engineering is simple: the rocket, the reactor, and the desalination plants are devices that are expensive to develop, to be sure, but their feasibility is relatively easy to assess; and their success is relatively easy to achieve, once one understands the scientific principles that underlie them.

It is therefore tempting to raise the following question: In view of the simplicity of technological engineering and the complexity of social engineering, to what extent can social problems be circumvented by reducing them to technological problems? Can we identify quick technological fixes for profound and almost infinitely complicated social problems, "fixes" that are within the grasp of modern technology and that would either eliminate the original social problem without requiring a change in the individual's social attitudes or would so alter the problem as to make its resolution more feasible? To paraphrase Ralph Nader, to what extent can technological *remedies* be found for social problems, without first having to remove the *causes* of the problem? It is in this sense that I ask, "Can technology replace social engineering?"

The Major Technological Fixes of the Past

To better explain what I have in mind, I shall describe how two of our profoundest social problems—poverty and war—have in some limited degree been solved by the technological fix rather than by the methods of social engineering. Let me begin with poverty.

The traditional Marxian view of poverty regarded our economic ills as being primarily a question of maldistribution of goods. The Marxist recipe for elimination of poverty, therefore, was to eliminate profit, in the erroneous belief that it was the loss of this relatively small increment from the worker's paycheck that kept him poverty-stricken. The Marxist dogma is typical of the approach of the social engineer: one tries to convince or coerce many people to forego their short-term profits in what is presumed to be the long-term interest of the society as a whole.

The Marxian view seems archaic in this age of mass production and automation, not only to us but apparently to many Eastern Bloc economists. For the brilliant advances in the technology of energy, of mass production, and of automation have created the affluent society. Technology has expanded our productive capacity so greatly that even though our distribution is still inefficient, and unfair by Marxian precepts, there is more than enough to go around. Technology has provided a "fix"—greatly expanded production of goods—that enables our capitalist society to achieve many of the aims of the Marxist social engineer without going through the social revolution Marx viewed as inevitable. Technology has converted the seemingly intractable social problem of widespread poverty into a relatively tractable one.

My second example is war. The traditional Christian position views war as primarily a moral issue: if people become good and model themselves after the Prince of Peace, they will live in peace. This doctrine is so deeply ingrained in the spirit of all civilized people that

I suppose it is blasphemy to point out that it has never worked very well—that people have not been good and that they are not paragons of virtue or even of reasonableness.

Though I realize it is a terribly presumptuous claim, I believe that Edward Teller may have supplied the nearest thing to a quick technological fix to the problem of war. The hydrogen bomb greatly increases the provocation necessary to lead to large-scale war—and not because people's motivations have been changed, not because they have become more tolerant and understanding, but rather because the appeal to the primitive instinct of self-preservation has been intensified far beyond anything we could have imagined before the H-bomb was invented. To point out these things today, with the United States in a shooting war, must sound hollow and unconvincing; yet the desperate and partial peace we have now is much better than a full-fledged exchange of thermonuclear weapons. One can't deny that the Soviet leaders now recognize the force of H-bombs, and that this has surely contributed to the less militant attitude of the U.S.S.R. And one can only hope that the Chinese leadership, as it acquires familiarity with H-bombs, will also become less militant. If I were to be asked who has given the world a more effective means of achieving peace—our great religious leaders, who urge us to love our neighbors and thus avoid fights, or our weapons technologists, who simply present us with no rational alternative to peace—I would vote for the weapons technologist. That the peace we get is at best terribly fragile I cannot deny; yet, as I shall explain, I think technology can help stabilize our imperfect and precarious peace.

The Technological Fixes of the Future

Are there other technological fixes on the horizon, other technologies that can reduce immensely complicated social questions to a matter of "engineering?" Are there new technologies that offer society ways of circumventing social problems and at the same time do *not* require individuals to renounce short-term advantage for long-term gain?

Probably the most important new technological fix is the intrauterine device for birth control. Before the IUD was invented, birth control demanded very strong motivation of countless individuals. Even with the pill, the individual's motivation had to be sustained day in and day out; should it flag even temporarily, the strong motivation of the previous month might go for naught. But the IUD, being a one-shot method, greatly reduces the individual motivation required to induce a social change. To be sure, the mother must be sufficiently motivated to accept the IUD in the first place, but, as experience in India already seems to show, it is much easier to persuade the Indian mother to accept the IUD once than it is to persuade her to take a pill every day. The IUD does not completely replace social engineering by technology; indeed, in some Spanish American cultures, where the husband's manliness is measured by the number of children he has, the IUD attacks only part of the problem. Yet in many other situations, as in India, the IUD so reduces the social component of the problem as to make an impossibly difficult social problem much less hopeless.

Let me turn now to problems that have from the beginning had both technical and social components—broadly, those concerned with conservation of our resources: our environment, our water, and our raw materials for production of the means of subsistence. The social issue here arises because many people by their individual acts cause shortages and thus create economic, and ultimately social, imbalance. For example, people use water wastefully, or they insist on moving to California because of its climate, and so we have water shortages; or too many people drive cars in Los Angeles with its curious meteorology, and so Los Angeles suffocates from smog.

The water resources issue is a particularly good example of a complicated problem with strong social and technological connotations. Our management of water resources in the past has been based largely on the ancient Roman device, the aqueduct: every water shortage was to be relieved by stealing water from someone else who, at the moment, didn't need the water or was too poor or too weak to prevent the theft. Southern California would steal from northern California, New York City from upstate New York, the farmer who could afford a cloud-seeder from the farmer who could not. The social engineer insists that such short-sighted expedients have gotten us into serious trouble; we have no water-resources policy, we waste water disgracefully, and, perhaps, in denying the ethic of thriftiness in using water, we have generally undermined our moral fiber. The social engineer, therefore, views such technological shenanigans as being short-sighted, if not downright immoral. Instead, he says, we should persuade or force people to use less water, or to stay in the cold Middle West, where water is plentiful, instead of migrating to California, where water is scarce.

The water technologist, on the other hand, views the social engineer's approach as rather impractical. To persuade people to use less water, to get along with expensive water, is difficult, time-consuming, and uncertain in the extreme. Moreover, say the technologists, what right does the water-resources expert have to insist that people use water less wastefully? Green lawns and clean cars and swimming pools are part of the good life, American style, 1966, and what right do we have to deny this luxury if there is some alternative to cutting down the water we use?

Here we have a sharp confrontation of the two ways of dealing with a complex social issue: the social engineering way, which asks people to behave more "reasonably," and the technologists' way, which tries to avoid changing people's habits or motivations. Even though I am a technologist, I have sympathy for the social engineer. I think we must use our water as efficiently as possible, that we ought to improve people's attitudes toward the use of water, and that everything that can be done to rationalize our water policy should be welcome. Yet, as a technologist, I believe I see ways of providing more water more cheaply than the social engineers may concede is possible.

I refer to the possibility of nuclear desalination.* The social engineer dismisses the technologist's simple-minded idea of solving a water shortage by transporting more water primarily because, in so doing, the water user steals water from someone else—possibly foreclosing the possibility of ultimately utilizing land now only sparsely settled. But surely water drawn from the sea deprives no one of his share of water. The whole issue is then a technological one: can fresh water be drawn from the sea cheaply enough to have a major impact on our chronically water-short areas, such as southern California?

I believe the answer is yes, though much hard technical work remains to be done. A large program to develop cheap methods of nuclear desalting has been undertaken by the United States, and I have little doubt that, within the next ten to twenty years, we shall see huge dual-purpose desalting plants springing up on many parched sea coasts of the world. At first these plants will produce water at municipal prices. But I believe, on the basis of research now in progress at Oak Ridge National Laboratory and elsewhere, that water from the sea at a cost acceptable for agriculture—less than ten cents per thousand gallons—is eventually in the cards. In short, I believe that, for areas close to the sea coasts, technology can provide water without requiring

*My discussion of nuclear desalination and very cheap nuclear power is of course out-of-date. Nuclear power is nowhere near as cheap as I had believed possible in 1966; however, as the technology matures, and particularly as reactors remain in service long after they have been amortized, these speculations, which seem so wrong in 1991, may yet prove to be correct.

a great and difficult-to-accomplish change in attitudes toward the utilization of water.

The technological fix for water is based on the availability of extremely cheap energy from very large nuclear reactors. What other social consequences can one foresee flowing from really cheap energy eventually available to every country regardless of its endowment of conventional resources? Though we now see only vaguely the outlines of the possibilities, it does seem likely that, from very cheap nuclear energy, we shall get hydrogen by electrolysis of water and thence the all-important ammonia fertilizer necessary to help feed the hungry of the world; we shall reduce metals without requiring coking coal; we shall even power automobiles with electricity, via fuel cells or storage batteries, thus reducing our world's dependence on crude oil, as well as eliminating our air pollution, insofar as it is caused by automobile exhaust or by the burning of fossil fuels. In short, the widespread availability of very cheap energy everywhere in the world ought to lead to an energy autarky in every country of the world and, eventually, to an autarky in the many staples of life that should flow from really cheap energy.

Will Technology Replace Social Engineering?

I hope these examples suggest how social problems can be circumvented, or at least reduced to less formidable proportions, by the application of the technological fix. The examples I have given do not strike me as being fanciful, nor are they at all exhaustive. I have not touched, for example, upon the extent to which really cheap computers and improved technology of communication can help improve elementary teaching without having first to improve our elementary teachers. Nor have I mentioned Ralph Nader's brilliant observation that a safer car, and even its development and adoption by the auto company, is a quicker and probably surer way to reduce traffic deaths than is a campaign to teach people to drive more carefully. Nor have I invoked some really fanciful technological fixes: such as providing air conditioners and free electricity to operate them for every Negro family in Watts on the assumption (suggested by Huntington) that race rioting is correlated with hot, humid weather; or the ultimate technological fix, Aldous Huxley's soma pills that eliminate human unhappiness without improving human relations in the usual sense.

My examples illustrate both the strength and the weakness of the technological fix for social problems. The technological fix accepts man's intrinsic shortcomings and circumvents them or capitalizes on them for socially useful ends. The fix is therefore eminently practical and, in the short term, relatively effective. One doesn't wait around trying to change people's minds: if people want more water, one gets them more water rather than requiring them to reduce their use of water; if people insist on driving autos while they are drunk, one provides safer autos that prevent injuries even in a severe accident.

But the technological solutions to social problems tend to be incomplete and metastable, to replace one social problem with another. Perhaps the best example of this instability is the peace imposed upon us by the H-bomb. Evidently the *pax hydrogenium* is metastable in two senses: in the short term, because the aggressor still enjoys such an advantage; in the long term, because the discrepancy between have and have-not nations must eventually be resolved if we are to have permanent peace. Yet, for these particular shortcomings, technology has something to offer. To the imbalance between offense and defense, technology says let us devise passive defense that redresses the balance. A world with H-bombs and adequate civil defense is less likely to lapse into thermonuclear war than a world with H-bombs alone, at least if one concedes that the danger of thermonuclear war mainly lies in the acts of irresponsible leaders. Anything that deters the irresponsible leader is a force for peace: a technologically sound civil defense would therefore help stabilize the balance of terror.

To the discrepancy between haves and have-nots, technology offers the nuclear-energy revolution, with its possibility of autarky for haves and have-nots alike. How this might work to stabilize our metastable thermonuclear peace is suggested by the possible political effect of the recently proposed Israeli desalting plant: the Arab states, I should think, would be much less concerned with destroying the Jordan River Project if the Israelis had a desalination plant in reserve that would nullify the effect of such action. In this connection, I think countries like ours can contribute very much. Our country will soon have to decide whether to continue to spend 5.5 billion dollars per year for space exploration after our lunar landing. Is it too outrageous to suggest that some of this money be devoted to building huge nuclear desalting complexes in the arid ocean rims of the troubled world? If the plants are powered with breeder reactors, the out-of-pocket costs, once the plants are built, should be low enough to make large-scale agriculture feasible in these areas. I estimate that for 4 billion dollars per year we could build enough desalting capacity to feed more than 10 million new mouths per year (provided we use agricultural methods that husband water), and we would thereby help stabilize the metastable, bomb-imposed balance of terror.

Yet I am afraid we technologists shall not satisfy our social engineers, who tell us that our technological fixes do not get to the heart of the problem; they are at best temporary expedients; they create new problems as they solve old ones; to put a technological fix into effect requires a positive social action. Eventually, social engineering, like the Supreme Court decision on desegregation, must be invoked to solve social problems. And of course our social engineers are right. Technology will never *replace* social engineering. But technology has provided, and will continue to provide, broader options to the social engineer, to make intractable social problems less intractable; perhaps, most of all, technology will buy time, that precious commodity that converts violent social revolution into acceptable social evolution.

Our country now recognizes, and is mobilizing around, the great social problems that corrupt and disfigure our human existence. It is natural that in this mobilization we should look first to the social engineer. But unfortunately the apparatus most readily available to the government, like the great federal laboratories, is technologically oriented, not socially oriented.* I believe we have a great opportunity here: for many of our seemingly social problems do admit of partial technological solutions. Our already deployed technological apparatus can contribute to the resolution of social questions. I plead therefore first for our government to deploy its laboratories, its hardware contractors, and its engineering universities around social problems. And I plead secondly for understanding and cooperation between technologist and social engineer. Even with all the help he can get from the technologist, the social engineer's problems are never really solved. It is only by cooperation between technologist and social engineer that we can hope to achieve what is the aim of all technologists and social engineers—a better society, and thereby a better life, for all of us.

*Today (1991) Oak Ridge National Laboratory employs several dozen social scientists—economists, geographers, anthropologists, and sociologists.

Part II: Scientific Administration

During most of my career, I have earned my livelihood supervising other scientists (that is, administering science) rather than doing science myself. As a sort of avocation, I began to write articles on what I have called the philosophy of scientific administration. This moonlighting career began when I became a member of President Eisenhower's Science Advisory Committee in 1959. I never became a real insider on PSAC. Perhaps because I was a bit out of the main stream of PSAC activities, my writings on science policy were philosophic rather than practical; yet I hoped these philosophic ruminations might be relevant to the issues facing PSAC. Most central, of course, was the allocation of resources between science and other activities and among competing scientific claimants. I first wrote on criteria for scientific choice for Edward Shils' journal *Minerva* in 1962; this article actually was based on a talk I gave to the University of Tennessee Chapter of $\phi\kappa\phi$ on the problems of big science (*Phi Kappa Phi Journal*, Vol. 421, Summer 1962, pp. 3–18). My early writings on this subject were published in my 1967 collection of essays, *Reflections on Big Science* (MIT Press, 1967).

Criteria for scientific choice has since become a cottage industry, one in which I occasionally participate. The first three essays in this group, "The Axiology of Science" (1970), "Unity As a Criterion of Scientific Choice" (1984), and "The Philosophy and Practice of Scientific Administration" (1988), pretty well represent my current views on the problem of scientific choice. The "Unity" paper was presented at the seventeenth International Conference on the Unity of the Sciences, a conference that I chaired in 1986, 1987, and 1991. I should mention that the very idea of explicit criteria of choice, and the implied intervention by government in the scientific enterprise, raised the hackles of the many disciples of economist Friederich Hayek who attended the conference. For Hayekians, tampering with the market is anathema, even where Big Science is concerned. I regard such staunch faith in the market as being a bit fanatical, but perhaps this is not the place to voice such seemingly socialist prejudice.

The last essay, on scientific information, is more practice than philosophy. I include it in this volume because, for a time, I was much involved in science information, both as Director of Oak Ridge National Laboratory and as Chairman of the PSAC Committee on Information; and also because this was my one really tangible contribution to PSAC during my two and a half year tenure.

The Axiology of Science

Peter Caws, in his book *The Philosophy of Science*, divides philosophy into four main branches: ontology, the theory of being; epistemology, the theory of knowledge; logic, the theory of inference; and axiology, including ethics and aesthetics, the theory of value. He then points out that the philosophy of science deals primarily with epistemology and logic, and a little with ontology. But axiology is, to quote Caws, "comparatively neglected . . . . Value in science . . . is . . . a matter of style—it involves a sense of proportion, and a feeling for the 'fit' of theory to the world."[1] Even this is an overstatement: to all intents, axiology is ignored by philosophers of science.

I make no claims to being a philosopher; and I am aware of A. Cornelius Benjamin's observation that "most scientists write bad philosophy and most philosophers write bad science "[2] Yet I cannot let questions of what *ought* to be in science, of what constitutes scientific taste and scientific value, be rejected so brusquely. I believe that axiology deserves a serious place in the philosophy of science, comparable, say to that of epistemology and logic; indeed, elements of an axiology of science already exist.

Who would be interested in an axiology of science? Perhaps I should ask, first, "Who is interested in the epistemology and logic of science (aside from the philosophers of science)?" The answer, according to the philosophers of science, is the practicing scientist: the program of philosophy of science is to clarify questions that arise in, but transcend, science. Now in all honesty I have never once seen a scientist doing something differently in his scientific work because of some relevant stricture or canon from the philosophy of science. And yet, despite

American Scientist **58** (6), 612–617 (1970). Reprinted by permission.

Peter Medawar's caustic assertion that "scientists treat the 'philosophy of science' with exasperated contempt,"[3] the philosophy of science *is* useful to the scientist: it makes him aware of his implicit assumptions, even if it is not a tool to be used explicitly like calculus or programming. In this sense, science is enriched by the philosophy of science.

Who would be interested in the axiology of science? Most obviously, the scientific administrator: here, I regard scientific administration in its fundamental aspect—that is, the allocation of scientific resources—rather than in its secondary, though important, aspect of scientific housekeeping. Scientific administration in this sense—whether it is conducted at the level of the individual scientist who decides to do this rather than that, or at the level of the research director who sets a whole laboratory on this course or that, or at the level of the Bureau of the Budget, which directs a whole nation's science policy one way or another—in every instance involves questions of value. Is it more important, or more valuable, or somehow better, to do this rather than that, to support high-energy physics rather than nuclear physics, or oceanography rather than space science? Again, I know of no case where the admittedly rudimentary existing axiology of science has provided a unique recipe for action for any scientific administrator. And yet, in much the same sense as the traditional philosophy of science helps clarify the underlying assumptions of the scientist, so I conceive the axiology of science to be useful to the scientific administrator.

I need not dwell at length on why this is a timely question. The dwindling of the scientific budget has put great pressure on scientific administrators at every level to make sharper, and often more painful, choices. If an axiology of science can be devised that will help the adminstrator make better choices, or at least give him some underpinning for his instincts, this would be useful.

Beyond this, an axiology of science provides a framework and a language for scientific administration, just as the philosophy of science provides such a framework for science itself. As background for this assertion, let me describe how scientific choices are actually made. Science operates, by and large, outside the ordinary market place. To be sure, there is Harvey Brooks' intellectual market place,[4] Polanyi's "Republic of Science,"[5] which sifts good science from bad science. The feedback from the intellectual market place works well *within* a particular scientific universe. Thus in 1963 the community of high-energy physicists was required to make a judgment as to whether higher energy (as embodied in the 200 GeV Batavia Accelerator) or higher intensity (as embodied in the MURA Fixed Frequency Alternating Gradient cyclotron [FFAG]) was the more important. The decision, arrived at largely through the working of the intellectual market place, went to higher energy. Batavia is under construction; FFAG is forgotten.

But, between scientific fields, it is hard to conceive how the intellectual market place can work. Space physicists don't communicate with biomedical scientists, nor do oceanographers communicate with chemists. In the absence of feedback from the intellectual marketplace, relative importance must be assessed and resources allocated by political processes. This is the way resources have always been allocated by government, the paradigm of a nonmarket economy. Just as ordinary politics profit by philosophic underpinning, by discussion of unstated assumptions and analysis of conflicting positions, so one would hope that scientific politics, the mechanism by which many of the largest scientific decisions are made, would be improved by a philosophic—or, more specifically, axiological—underpinning. We would not hope or want to replace politics with philosophy; rather the philosophic discourse would provide a language for, and an approach to, the political dialog.

There is another, quite different, application of axiology of science—the formulation of the scientific curriculum. For what is curriculum building if not an implicit ordering of importance in science? Why do we today believe to be so quaint Professor Grant's 1840 examination questions in comparative anatomy that begin: "By what special structures are bats enabled to fly through the air? and how do the galeopitheci, the pteromys, the petaurus and the petauristae support themselves in the light element? . . . and explain the structures by which the cobra expands its neck . . . and how do flying fishes support themselves in air?"[6] And why do we consider codons and operons and transfer RNA somehow better science and much worthier of concern for students of science than Professor Grant's lovely flying fish? Both are science; yet, by current standards, one is "better" science than the other. This implies an axiology of science: can we make this axiology more explicit?

Finally, axiological issues are implicit in the very practice of science. Every scientific discovery is judged first on epistemological and logical grounds. Is this piece of science correct and valid? Does it hang together logically? Are there epistemological flaws in it? Most published science passes these tests tolerably well. There then remain axiological questions: of two pieces of science that are equally correct, equally valid, how does one decide which is more important, which "better?" We usually say, ultimately, it's a matter of taste. Einstein spoke of beauty in a scientific theory, Popper of simplicity, Medawar of explanatory power, clarifying power, and originality. These are all elements of scientific taste. I have nothing to add to these judgments, which are applied by one scientist in evaluating another's work. Most of what I have to say applies to the axiology of science as it relates to the administration of science on the one hand or its teaching on the other.

Implicit Axiological Attitudes Toward Science

Several very broad attitudes toward different styles of science are so deeply a part of the scientist's prejudices as hardly to be recognized as implying an axiology. These I call implicit axiological attitudes toward science. I would include in this category such homilies as: Pure is better than applied; General is better than particular; Search is better than codification; and Paradigm breaking is better than spectroscopy.

Pure is better than applied

Attitudes toward the relative worth of pure and applied science go back at least to Francis Bacon and Thomas Sprat. It was Bacon who observed that science was done because from science "we learn how to make two blades of grass grow where one grew before." According to Medawar, in the early days of the Royal Society, science with no thought as to its immediate usefulness was dismissed as play. A change occurred during the Romantic Era. In England by the middle 1850s pure science began to be valued above applied science; this bias persists and, if anything, has become stronger down to the present.

Medawar suggests that the English taste for purity in science probably derives from "the conscious and deliberate perpetuation by our public schools of the Platonic conception of activities that did or did not become a gentleman." According to Medawar, "it was put admirably by John Gillies in the introduction to his translation of *Aristotle's Ethics and Politics* (London, 1797). Aristotle's experimentations . . . were 'confined to teaching nature in the fact, without attempting, after the modern fashion, to put her to torture'; for philosophers . . . to bend over a furnace, inhaling noxious steams; to torture animals, or to touch dead bodies, appeared to them operations . . . unsuitable to their dignity."[7]

I suppose Americans would attribute the current taste for pure as opposed to applied science to the social organization of the university.

The university is the preeminent intellectual institution of our society. It is disciplinary, in contrast to the mission-oriented institute, which is interdisciplinary. Insofar as pure science tends to be disciplinary, it is natural that the university should value the pure above the applied. Moreover, the taste of the university, because of the university's intellectual preeminence, becomes the taste of society.

Beyond this, pure science tends to be internally generated, arising from the logic immanent in science itself; applied science arises from needs that lie outside science, or perhaps outside a given discipline. It is natural for the university professor to gravitate toward pure science, since the pursuit of pure science by and large involves little outside sanction. The pure scientist sets his own problem, and this is congenial to the university professor, who almost by definition goes where his intellectual instinct directs.

Pure science, because its problems are set from within, is in a sense easier than applied science, and this may also help explain its popularity. (Of course, this is balanced by the limited aims of applied science—to "cure" rather than to "understand.") A theory of turbulence would be a fine thing because turbulence is important to much of engineering. But a satisfactory theory of turbulence has evaded hydrodynamicists: instead, they work on soluble problems whose connection with the practical phenomenon of turbulence is often relatively weak. A cure for cancer is very desirable but very difficult: so cancer researchers work on problems in "pure" biochemistry, protein synthesis, and the like, which are soluble but whose relevance to cancer is always conjectural until proved otherwise.

It is quite possible that the new cry of relevance within the universities may change our relative valuation of the pure and the applied. Already we see a turning away, especially by younger scientists, from science that is remote and detached and a concern—even a yearning—for science that is relevant. Whether or not this shift occurs (taking us back perhaps to the attitude of the late seventeenth century, when applied science was valued above pure), the important point is that society's, and even the scientist's, valuation derives from social pressures that lie outside science.

General is better than particular

A second example of an implicitly axiological question in science is the relative valuation of the general and the particular. Most scientists would hold the general to be better than the particular. This view was presented most maturely by Eugene Wigner in his Nobel lecture, "Events, Laws of Nature, and Invariance Principles,"[8] though Karl Popper, in his *Logic of Scientific Discovery*,[9] had already spoken of the relation between events and laws of nature. As Wigner put it, our observable world is made up of individual events: falling apples, particular flying fish, and so on. The aim of science is to subsume these individual events in laws of nature. We can predict the course of every falling apple if we know Newton's laws and if we know the initial and boundary conditions. Wigner, the inventor of symmetry principles, goes further and shows the remarkable extent to which at least some of the most important laws of nature—conservation of energy and of momentum, for example—follow from the most rudimentary symmetry properties of space-time (conservation of energy from invariance with respect to time displacement, conservation of momentum from invariance with respect to rotations in space-time).

An ordering such as this already can be described as a rudimentary axiology: symmetry principles are more fundamental, more important, "better," if one likes, than laws of nature; and laws of nature are more important and "better" than events. This ordering derives from the principle of scientific parsimony, which is usually accepted as one of the goals of science: to explain as much as possible with as little as possible. Unlike other goals of science—in particular, its usefulness—this goal arises from within science itself. It is hard to quarrel with this goal; but it is also hard to give it a fully logical justification.

The principle of scientific parsimony and the consequent striving toward generality has an important practical justification: it helps scientists cope with the information explosion. Science seeks always to find more general principles—Wigner's laws of nature—from which events may be derived. There are great practical advantages here since, once a general principle is established, it is no longer so important to keep the events in mind. The events, to use Medawar's felicitous phrase, are annihilated by the general principle.

To every modern scientist, science that deals with disparate, unconnected facts is "poorer" science than science that deals with powerful generalizations. And yet one can overdo this love of the general as opposed to the particular. For every general law implies particular instances only in principle. And to know in principle is not the same as knowing in fact and in detail. From quantum chemistry we can predict, in principle, the properties of a nucleic acid; yet to do so in practice would be very difficult. As we go to more powerful general principles, we gain in breadth of outlook, but we lose in resolution. The genetic code may imply all of developmental biology; but the essence of developmental biology—for example, the details of how eye cells of the newt transform into a crystalline lens—is surely, in practice, far beyond the competence of the genetic code.

To some extent, what is at issue here is the possibility of reducing a set of related events to a science. The events that constitute natural science by and large have enough underlying regularity to admit useful generalizations. The events follow from the laws of nature though, as the events become complicated, this becomes more a matter of "follows in principle" than "follows in fact." But what about the social sciences, and more particularly those social sciences like sociology and political science, where one no longer makes much pretense of predicting events from general principles? Here the relative merit of general and particular is reversed from the usual order, although, because social scientists tend to borrow paradigms from the natural sciences, I suspect that they would uphold the superiority of the general.

Again, my purpose here is not to argue that the general is better than the particular or the converse. It is rather to show this as an axiological judgment, one that in this instance can be traced to the principle of scientific parsimony. Obviously our love for the general as opposed to the particular would not be so ardent if we were less taken by the law of scientific parsimony and more taken, say, with the applicability of science to the problem of pollution. This is meant only to show the close, and hardly surprising, connection between axiological questions in science and the overall aims we set for science.

Search is better than codification

I turn now to the distinction between science as search for new knowledge and science as the codifier of existing knowledge. Most scientists would admit that both aspects of science are valid, that Newton's laws and the second law of thermodynamics will in some approximation remain true forever and will always remain part of science.

This is in opposition to views expressed by some of our modern curriculum reformers. "Scientific knowledge is revisionary," says Joseph J. Schwab. "It is a temporary codex."[10] Yet this is surely an overstatement; science does possess a consolidated corpus, much of which does *not* change. The boiling point of uranium is a datum of science, and, once accurately measured, it can change only in insignificant degree.

Whether one deems science as search or science as codifier to be the more important depends on what one expects to do with science. To one who values science for the control it gives us over nature—that is, broadly, to the engineer, or physician, or other scientifically based professional—science is largely "codified" science. For, as Derek Price has stressed, technology tends by and large to utilize the older, better codified parts of science.[11] On the other hand, to the practicing scientist who is trying mainly to extend human understanding, science is scientific research.

Scientists often get around the necessity of making this judgment explicitly by relegating the codified or closed part of science to engineering. Reactor physics is now part of nuclear engineering; classical mechanics is part of mechanical engineering. But herein lies a danger for the development of science. True, the most advanced researches in a narrow field of science may often be pursued without much contact from the well-codified, older parts of the field. Yet the older parts of science to a strong degree remain relevant for neighboring scientific disciplines. Thus the most important discovery in molecular biology, the double helix of DNA, used relatively old techniques of the x ray crystallographer.

My point is not primarily to remind our curriculum reformers that they must not reject science as codifier in favor of science as search. It is rather that such a judgment already is heavily axiological in implication. The basis for this axiological judgment comes, again, from outside science, from our original assumption about the ultimate goal of science: means for control or means for understanding.

Paradigm breaking is better than spectroscopy

I use paradigm breaking in the sense of T. Kuhn[12]: scientific "progress" is punctuated by "revolutions," which break existing paradigms; in between such revolutions, scientists add details within an existing paradigm—this I call "spectroscopy." Thus the discovery of the first xenon compound by Neil Bartlett broke the paradigm; the massive collection of many new noble gas compounds, largely at Argonne, was spectroscopy. Kuhn's axiology, though implicit, is clear: the breaking of paradigms is "better" science than the practice of spectroscopy.

One can hardly quarrel with Kuhn's judgment; yet, if one took his axiology too seriously, little would remain that was worthwhile. Most science is spectroscopic. The breaks in the paradigm come seldom, and one simply cannot wait for them.

Some scientific fields acquire so strongly specialized a spectroscopic character as to elicit bewilderment and even haughty criticism from scientists in other fields. To an extent, this may be happening to nuclear physics, a field that is now fifty years old (Rutherford disintegrated nitrogen with alpha particles in 1919). Yet, to the nuclear physicist, the spectroscopy remains exciting and vital. He argues, and with considerable justification, that out of his spectroscopy come breaks in the paradigm.

Again we come to the goals set by the scientist. In the case of the nuclear physicist, if we should ask him "What are the goals of your science?" he probably would say "To understand nuclear structure." But this is a rather diffuse aim, of quite different character than, say, the aim of controlled-fusion research. We would know when we have achieved controlled fusion; I doubt that we shall know when we have "understood" nuclear structure. Thus, once we have achieved controlled fusion, the importance of the basic research that led to this achievement might be expected to diminish. On the other hand, spectroscopic exploration of nuclei has no obvious end point. Since we probably will not know when we have "understood" nuclear structure, I suppose we will not know when nuclear spectroscopy (in its broad sense) becomes irrelevant to our underlying aim.

What actually happens, of course, is that a viable spectroscopy is punctuated by breaks in the paradigm. Thus in nuclear spectroscopy we have, over the past ten years, discovered isobaric analog states, the double hump in fission, the possibility of super-heavy elements—each of which represents a small break in the paradigm. Eventually paradigms even of this sort and weight will cease to be broken: it is only at that time that one could expect physicists to lose interest in nuclear structure and for the rest of physics to relegate the subject to a secondary position.

Criteria for Scientific Choice

In a different category from the implicit axiologies are the explicit attempts to order specific fields of science or specific scientific enterprises. Here one asks the explicit question: Can one construct a set of *a priori* criteria according to which one can rate various scientific enterprises or scientific fields? The aim here is not philosophic but practical. It is to give to the scientific administrator some guidance for allocating his resources. And, indeed, the literature on this subject comes mostly from administrators who themselves are trying to figure out how they should divide their resources among competing scientific claimants. Let me illustrate this approach by describing the "criteria for scientific choice" that I first proposed in 1962.[13] In this axiology, I propose two different kinds of criteria, internal and external, according to which scientific activities can be valued. The internal criteria arise from within the science itself, or from its social structure and organization; the external criteria stem from the social or other setting in which the science is embedded.

Among the internal criteria, I identify ripeness for exploitation and caliber of the practitioners.

Ripeness for exploitation

This is perhaps the most important of all internal criteria: science is the art of the soluble. Thus, one must ask of any scientific activity whether it is at a stage where effort on it is likely to produce results. It would have been foolish to work on nuclear energy before fission was discovered.

Caliber of the practitioners

Are the people in the field good, and do they know what they want to do? This is a particularly hard judgment; every scientific community has its giants, but a giant in one community might be a pygmy in another. Such disparity of standards plagues every university dean, every research director. If he himself is a theoretical physicist, he will wonder why his engineers are duller than his theoretical physicists; or, if he is an engineer, he will wonder why his theoretical physicists are so impractical. From my own experience, I think one can make judgments about people and about the field they represent. The really great scientists I have known have, with rather few exceptions, been able to explain what they are up to in simple language when this was necessary. They have the knack of knowing—and always keeping uppermost—the essence of what they are doing, and what its relation is to other things. To the extent that practitioners in a field can articulate their aspirations and plans, they and their field deserve serious attention; to the extent that they cannot, they deserve little support.

Thus, for valuing scientific activities, one can propose criteria that arise from the intrinsic logic or social structure of science. But, with respect to the largest judgments that society is called upon to make regarding science, whether to explore the Moon or to screen thousands of potential chemotherapeutic agents, the more important criteria must arise from outside science or from outside a given field of science. Such criteria, which I call external, ask of a given scientific activity that it matter to someone outside the scientific community in which the activity is undertaken—to engineers, to other scientists, to the society at large.

I have identified three such external criteria: technological merit, social merit, and scientific merit.[14] The first two are almost self-explanatory. Technological merit means the degree to which a scientific activity contributes to achievement of a technological aim: accurate measurement of the number of neutrons produced per fission is necessary before we can design a breeder reactor. Social merit means the degree to which a scientific activity contributes to achievement of a social goal: the moon shot obviously created a great feeling of international unity, as well as enhancing the prestige of the United States.

The establishment of social goals is of course a difficult task that has bedeviled politicians and philosophers. I have been criticized by R. Gotesky for assuming, in my axiology of science, the social goals as given.[15] Gotesky holds that, since scientific institutions are clients of states, the only social goals that science does contribute toward achieving are those that a particular state believes to be in its own interest. So extreme a view seems to me to be beside the point: there are many scientific goals, such as curing cancer or achieving controlled fusion, that are indeed in the interest of mankind rather than of any particular state.

Nevertheless, there is a very important related issue that has long troubled scientists, and that is now being raised with much urgency, especially by the counterculture. Increasingly, we hear questions raised about whether science ought to investigate certain areas if the findings might have deleterious social consequences. One such question is the predetermination of a baby's sex. Another is the correlation of race with intelligence. Answers to such questions, if they can ever be reached, obviously raise issues that lie outside science.

I use the term "scientific merit" in the following specialized sense: the scientific merit of a field of science is to be measured by the degree to which it contributes to and illuminates the neighboring fields in which it is embedded. In a way, such a criterion of merit derives from the law of scientific parsimony, since this criterion implies that one aim of science is to unify our picture of the world; a scientific activity that contributes to the unification is to be valued over one that does not. A scientific activity that bears most broadly, that illuminates most widely, that *matters* to the largest number of scientists is somehow "better" than one that does not. Or, again, one can view such an axiology as a kind of scientific utilitarianism. What is best in science is what is best for the largest number of scientists. If one is concerned with enhancing the totality of scientific progress, then those elements of science that contribute most in getting on with the whole job are surely better than those that do not contribute as much toward this end.

Before turning to weaknesses and criticisms of the axiologies I have described, I should like to add another criterion that I think deserves attention; at least, I have on occasion applied it in my own scientific administration. I have in mind the obverse of the criterion of scientific merit; it might be called "derived" scientific merit and would be stated thus: "The derived scientific merit of scientific activity is to be measured by the degree to which the activity is influenced and is illuminated by other scientific fields." Important advances in science often occur when a technique from a different field is brought in to help look at an old question in a new way. To give some examples: the application of group theory to quantum mechanics, or the application of nuclear magnetic resonance to organic analysis, or the application of carbon dating to archaeology. The scientific administrator's heart is warmed by a proposal to look at a perplexing question with a tool or set of tools derived from some entirely unrelated parts of science. To the extent that this phenomenon can be generalized to big branches of science (for example, the revolution in some behavioral sciences precipitated by the computer), I would propose elevating this criterion to an axiological principle.

The attempts to establish criteria of choice have suffered from the disadvantage, uncommon in philosophy, of having rather specific practical applications. Thus Michael Gibbons claims that the recommendation by the British Council for Scientific Policy to support the CERN 300 GeV accelerator was much influenced by criteria of this sort.[16] I have tried to rate specific scientific fields according to these criteria and, in consequence, have incurred the wrath of those high-energy physicists and behavioral scientists who believe I undervalue their fields. Some of the critics have themselves contributed alternative axiological principles, and I shall describe some of them.

Victor Weisskopf argues that the criterion of scientific merit is too restrictive. To be sure, science that illuminates other fields is meritorious. But, as Weisskopf puts it, science progresses along two dimensions—an "extensive" dimension and an "intensive" dimension.[17] Extensive science seeks to apply known principles, to explore ever more complicated situations. It is science concerned with boundary conditions, with applications of known fundamental laws to new situations. Extensive science includes chemistry, solid-state physics, most of biology (unless we stumble upon "biotonic" laws). Of extensive science, we can properly ask that it bear on other branches of science or on other affairs of man: the mere collection and cataloging, or even organization, of specific instances are not sufficient justification for large-scale support. Such science properly ought to matter to someone other than the scientist doing the science.

The other kind of science, "intensive" science, seeks out totally new and fundamental laws of nature. In physics, Weisskopf would elevate only high-energy physics and cosmology to this status. Since the aims of these sciences are deep and narrow, it would be unfair to require them to bear on other parts of science—unfair perhaps, but not irrelevant. If the science under question is cheap and demands little of the society, then fine: let it proceed on its pristine path. But if the science is very expensive, then society can properly ask, "Why the rush? Is it really necessary to follow the 200 GeV immediately with a 1,000 GeV accelerator, or cannot one wait until new results are digested before taking the next step?"

The other criticism of *a priori* criteria of choice denies the possibility of such criteria: science is unpredictable; its influence on future developments is impossible to assess. Karl Pearson, only a few years before Marconi, pointed to Hertzian waves as examples of discoveries with no possible use; and Hardy retreated to number theory because it was pure, only to learn later that Bohr and Kalckar applied Hardy and Ramanujan's theorem on partitioning of integers to the distribution of nuclear energy levels. Yet, to my mind, such criticism misses the point. When we have enough for all—as when we are speaking of little science—choices rarely have to be made so explicitly, so sharply. It is only when the choices must be made on the largest scale that *a priori* judgments must be made.

Now, in practice, these largest judgments will surely have strong political overtones in their making. Imperfect as criteria for choice must be, they are still better than no criteria. A politics without an underlying axiology is in danger of becoming venal. So even though the usefulness or applicability of a scientific activity is difficult to assess, it is still worth *trying* to make the assessment.

In conclusion, I find that axiology of science has many dimensions: *a priori* criteria of choice, *a posteriori* criteria for *posteriori* evaluation, implicit elements. Some of the criteria seem to me to be embedded in the underlying strategy of science—for example, the principle of parsimony. Others come from our conceptions of other goals of science—science for use, science for understanding. Still others come from the social structure and organization of science. I do not pretend to have given a whole analysis of scientific taste, or of the philosophy of scientific administration. Such questions are what Cornelius Benjamin calls "speculative questions," meaning I suppose that one can much easier raise than answer them. Perhaps my question will prompt philosophers of science to give to axiology of science the attention it deserves; future administrators of science, if not scientists, will be grateful for any wisdom that this dialog engenders.

References

[1] P. Caws, *The Philosophy of Science* (D. Van Nostrand, Princeton, NJ, 1965), pp. 13, 267, 334.

[2] A. C. Benjamin, *An Introduction to the Philosophy of Science* (MacMillan, New York, 1937), p. 36.

[3] P. B. Medawar, *The Art of the Soluble* (Methuen, London, 1967), p. 151.

[4] H. Brooks, *The Government of Science* (MIT Press, Cambridge, MA, 1968).

[5] M. Polanyi, Minerva **1**, 54–73 (1962).

[6] Medawar, *op. cit.*, p. 114.

[7] *Ibid.*, p. 13.

[8] E. Wigner, Science **145**, 995–99 (1964).

[9] K. R. Popper, *The Logic of Scientific Discovery* (Hutchinson, London, 1959).

[10] Biological Sciences Curriculum Study, Joseph J. Schwab, Supervisor. 1964. *Biology Teachers' Handbook.* New York: John Wiley.

[11] D. J. de Solla Price, Technology and Culture **4**, 564 (1965).

[12] T. S. Kuhn, *The Structure of Scientific Revolutions* (University of Chicago Press, 1962).

[13] A. M. Weinberg, Minerva **1**, 159–71 (1963).

[14] *Ibid.*, p. 164.

[15] R. Gotesky, Journal of Value Inquiry **3**, 298 (1969).

[16] M. Gibbons, Minerva **8**, 181–91 (1970).

[17] V. F. Weisskopf, Science **147**, 1552–54 (1965).

Unity As a Criterion of Scientific Choice

Scientific activity includes both its administration and its practice. By administration of science I mean not the housekeeping of science but rather the art of choosing, among the infinitely many possible questions answerable by science, which questions to ask. By the practice of science I mean the actual conduct of research: theorizing, observation, measurement, interpretation of results, communication of results. Put otherwise, administration is concerned with what to do, practice is concerned with how to do it; or, with less accuracy, administration is—roughly—strategy, practice is tactics.

This distinction between administration and scientific practice holds at every level. The individual scientist must decide which research he ought to carry out next; he must then carry out the research. He is therefore both a scientific administrator and a scientific practitioner. A scientist's proficiency as an administrator is a measure of his scientific taste—for what is scientific taste but the knack of choosing worthwhile problems? The administrative facet of a scientist's work, his "taste," rarely intrudes explicitly—good scientists have it, poor ones do not.

Some scientists excel as administrators, others as practitioners. Thus, James Conant, in comparing the scientific styles of Lavoisier and Priestley, said:

> Lavoisier's lasting contribution was made because he placed his experiments in the framework of an ambitious attempt to explain a great many facts in terms of a grand conceptual scheme. It would not be too

First published in Minerva **22**(1), pp. 1–12 (1984). Reprinted by permission.

> misleading to call him a master strategist in science. Priestley on the other hand probably excelled Lavoisier as an experimenter but he failed to appreciate fully the significance of his results in terms of the great question of the day—combustion and calcination . . . he was a great tactician, but a poor strategist.[1]

The individual scientist working at his bench—today more likely to be at his Apple computer—epitomizes "little science." In "little science," administrator and practitioner are the same person. As the size and complexity of the questions addressed increase—that is, as the science becomes "big science"—the split between administrator and practitioner becomes more pronounced. Since much more is at stake in "big science" than in "little science," the strategic choices must be made much more explicitly and self-consciously in the former. The director of a large laboratory must, at least in theory, devote most of his time to choosing between competing claimants on his always limited budget. At the highest level of scientific activity—that is, the allocation of a country's total scientific resources between, say, high-energy physics, molecular biology, and environmental science—and the carrying out of national policy, the separation between administration and practice is practically complete. The President's science adviser spends all his time worrying about allocation; he has no time left over for the details of how the "science pie" he has cut is actually eaten.

Scientific values underlie criteria of choice according to which we decide upon the "worth" or "validity" of scientific activities. Values underlie both the administration and the practice of science: these values therefore constitute the subject matter of metascience. The values that underlie the practice of science and those that underlie the administration of science are different. Corresponding to the separation of science into its two aspects, practice and administration, there are two separate sets of metascientific values—one for scientific practice, another for scientific administration. In speaking of values in science, I shall therefore deal with the two sets of values separately.

The Values of Scientific Practice

The primary question asked of every scientific discovery is "Is this discovery true?" Indeed, since science is usually regarded as a search for truth, truth is the criterion by which every scientific assertion is judged, and it must therefore be regarded as the underlying value of scientific practice.

I speak deliberately of scientific practice, not of science. This is an essential point, since we know that two scientific discoveries may be equally true—equally valid as judged by the criterion of truth—yet the one may be far more "significant," "worthwhile," or "valuable" than the other. In applying the criterion of truth, we are considering not whether the question purportedly answered by our research was a good or useful or important question; we are simply asking whether the question was answered correctly and convincingly. In short, truth is a value, a criterion of choice, only for the practice of science, not for the administration of science.

The philosophy of science, insofar as I understand it, is preoccupied primarily with epistemology and logic—that is, in establishing a basis for deciding whether or not a scientific discovery is true. It must, therefore, be regarded as the philosophy of scientific practice, since it deals hardly at all with the other aspect of science—namely, its "administration."

Many authors have been much intrigued by the notion of truth as a value, perhaps the primary value, in science. Among them are Jacob Bronowski,[2] Anatol Rapoport,[3] and Abraham Maslow.[4] My observation that truth is a value that underlies only the practice of science, not its administration, should not be taken as a disparagement of these authors. Science certainly aims for truth; and truth must therefore be regarded as a value of science, though it cannot be regarded as the only value.

But even truth cannot always be regarded as a fully operable or applicable criterion of merit for the practice of science. There are many important questions that are isomorphic with bona fide scientific questions and might therefore be regarded as scientific, but which, in principle, cannot be answered by science. I have called these questions "trans-scientific," since they transcend the capacity of science. Examples of such trans-scientific questions are the prediction of extremely rare events, or the prediction of trajectories for nonlinear systems whose behavior is extremely sensitive to small changes in initial conditions—so-called Poincaré instabities. Among rare events, perhaps the most relevant are those concerned with the response of animals, whether mice or human beings, to extremely small physical or chemical insults. We know that, on average, about 100 rems of radiation delivered to each of 1,000 mice is very likely to double the mutation frequency in each of the mice. What we cannot say, and possibly can never say, is whether 1 millirem delivered to each of 100 million mice, for the same 100,000 mouse-rem exposure, will also cause the same aggregate number of mutations. Thus, the two questions are isomorphic, yet the one involving the rare event is trans-scientific, the other is scientific.

As an example of the other kind of trans-scientific question, I mention attempts to predict the future, whether of the economy, of the demand for energy, or of the climate—where the underlying phenomena are so complicated as to be subject to Poincaré instabilities—or simply to demand more knowledge than we now possess. It was my frustration with these extremely important but all-but-unanswerable questions that led me to characterize most of the social sciences, with the possible exception of economics, as being "trans-scientific."

My separation of science from trans-science is probably too extreme, too schematic. The line between the two is more indefinite than I have implied. Many, if not most, of the problems that evoke violent reactions in public opinion—like the existence or nonexistence of a threshold for biological damage from exposure to toxic agents, or the biological basis for racial differences in intelligence, or the existence of the so-called nuclear winter—possess genuinely scientific as well as trans-scientific components. Thus, one finds scientists arguing vehemently regarding the capacity of science to give answers to these questions; the arguments are generally influenced by the social and political values of the antagonists. Such controversy, in which a scientist's commitment to truth is affected or replaced by his social and political values, must be contrasted with controversy over genuinely scientific issues, which can be settled by scientific methods and where social and political values intrude far less. The struggle between Mach and Ostwald, on the one hand, and Boltzmann, on the other, about the reality of atoms was often bitter, but I do not think the substantive positions of the antagonists were influenced by anything more than their commitment to scientific ideals. On the other hand, some recent arguments about the genetic basis of intelligence, or even the effects of extremely low levels of radiation, have surely been very much influenced by the political or moral attitudes of the adversaries.

None of this casts doubt on the primacy of truth as a criterion of merit in genuinely scientific, as opposed to trans-scientific, practice. All scientific practitioners profess a commitment to truth, and most of them adhere to it in scientific matters. Their moral or political values encroach on, or supersede, truth only where the latter is an insufficient criterion of validity, and that is in trans-science, not in science.

The Values of Scientific Administration

Scientific administration, in the narrow sense in which I use the term, asks not "Is this scientific proposition true?" but rather "Of two equally true scientific propositions, which is more worthwhile?" Both propositions may be equally valid, as measured by the criterion of truth, but one might be regarded as being more important than the other. The discovery of fission in uranium-235, and the discovery of a new energy level in the ^{235}U nucleus, are equally true; the former is obviously much more important than the latter. How do we know that one is more important than the other? Or, in the administrator's terms, how can we establish a schedule of priority between competing scientific projects?

Such after-the-fact judgments of the relative importance of different scientific discoveries of course have always been an intrinsic part of science. They give science an internal hierarchical structure that scientists find to be proper, to say the least. These judgments are also the very essence of scientific administration. Every administrator, at whatever level, is always deciding which scientific project to support and which scientific project not to support. Unfortunately, he must make the judgments before, not after, the science is practiced, and this requirement has given rise to a search for criteria of scientific choice. The debate on scientific choice, which began nearly a quarter of a century ago, has attracted considerable attention, especially among those formulating national scientific policy. This is an instance of a philosophic question—how to judge the relative value of competing scientific activities—that, at least in principle, has urgent practical application.

The debate itself was greatly encouraged by the editor of *Minerva*; most of the more theoretical writings on the subject have appeared in that journal. The modern debate began with Michael Polanyi's famous paper, "The Republic of Science: Its Political and Economic Theory," published in 1962,[5] though the question of how scientists choose among possible research projects was raised in a different idiom and with a different intellectual interest as early as 1939 by Robert K. Merton.[6] Professor Merton insisted that values arising outside science affected such choices:

> The foci of scientific interest are determined by social forces as well as the immanent development of science. We must therefore examine extra-scientific influences in order to comprehend . . . why scientists applied themselves to one field of investigation rather than another.[7]

I contributed to the debate of the 1960s in *Minerva* with two papers, both entitled "Criteria for Scientific Choice" (this title was suggested to me by Professor Shils). Other papers in the *Minerva* series were by Bruce Williams, Charles Carter, Stephen Toulmin, and Simon Rottenberg.[8]

The sharpest difference in outlook was between Polanyi and me. He regarded the republic of science as being governed by a free market in which the direction of scientific development resulted from the interplay of innumerable, decentralized decisions made by a myriad of individual scientific "administrators." He thus regarded science as a self-organizing structure guided by an "intellectual marketplace"—to use Harvey Brooks's phrase. He said that scientific research should be conducted as an unplanned competition among different scientific activities, each claiming greater scientific worth than the others and attempting to support its claim by the acceptance of its results by the scientific community according to scientific standards of truthfulness and significance. I, on the other hand, argued that, although little science could progress, and indeed had progressed, without explicit planning, the course of at least big science could be, and was being, affected by more central decisions based on "criteria of scientific choice."

As I have described in the previous essay, the criteria of choice that I thought should be applied by scientific administrators were of two kinds, internal and external. Internal criteria arise from within the administration of science and attempt to answer such questions as "Are the scientists competent?" and "Is this field at a point where progress can be expected—that is, is it ripe for exploitation?" If such internal criteria are not met, then the effort is likely to be fruitless—in other words, the internal criteria are criteria of "efficiency," since they are aimed at judging how likely it is that a given allocation of resources will actually yield positive scientific results. The underlying value here is "efficiency" in the use of resources to produce truthful and significant scientific results.

The external criteria arise from outside science or from outside the specific field being judged. Whereas the internal criteria are intended to test whether a proposed scientific activity is likely to be conducted efficiently in scientific respects, the external criteria are intended to test whether the activity is likely to be judged important, or useful, or worthwhile. The worth or value, as opposed to the truth, of an activity or proposition cannot be judged except from outside the given universe of discourse. Thus, to judge the worthwhileness of a given scientific project, we must go outside the project itself. I therefore proposed three external criteria of merit that arise from the outside: technological merit, social merit, and scientific merit.

By technological merit, I meant the technological relevance or usefulness of a scientific activity: for example, research on high-temperature plasmas obviously has great technological merit since it might lead to controlled-fusion energy, though some might have claimed, in the early days of the research on fusion, that useful points of departure were lacking.

By social merit, I meant the direct social impact of a scientific activity. For example, high-energy physics, largely conducted as an international collaborative enterprise, plays a role in furthering international understanding. Or, again, economics might be rated highly in social merit if it could find a way to combine full employment, economic growth, and stable prices.

Both these external criteria arise from outside science; they are mainly relevant to applied or rather "applicable" science. The value underlying these criteria is practical utility; if we apply these criteria, we choose to support those projects of scientific research that are socially or technologically useful. My third external criterion of merit, by contrast, arises from within science but outside the scientific field of activity under scrutiny; it alone is relevant to "pure" science. It was suggested to me by John von Neumann's beautiful statement about the necessity of a pure mathematical discipline, if it is to avoid fragmenting into a mass of incoherent detail, to return regularly to its antecedents in earlier, more classical branches of mathematics. I think that the same consideration ought, *mutatis mutandis*, to apply to empirical science. I therefore defined "scientific merit" of a given pure scientific activity in the statement: The scientific merit of an activity in pure science is to be measured by the degree to which it "contributes most heavily to and illuminates most brightly" the neighboring scientific disciplines in which the activity is embedded.

Unity As a Value in Administration of Pure Science

The internal criteria are criteria of efficiency in the attainment of scientific objectives. The external criteria of social and technological merit, often overriding in practice, are criteria of utility. They estimate the practical usefulness of a scientific discovery, but they affect the structure of science only in an incidental way. When we measure neutron cross sections in order to build better nuclear reactors, we hardly worry about rounding out our general picture of nuclear matter—we measure only those cross sections relevant to the technological task, whether or not these particular cross sections are of intrinsic scientific importance to the rounding out of our picture of the nucleus. Thus, one can speak of efficiency and utility as values underlying these criteria of choice.

Of the external criteria, I would regard the criterion of scientific merit as occupying the most fundamental theoretical position. It alone arises from within the whole of science and is intended to affect the underlying structure of science—the relation between the parts of science and each other. It alone among the criteria of choice concerns itself with the attainment of an orderly structure of science. This criterion judges the merit of a pure scientific activity by the illumination that its results might throw on the neighboring fields in which it is embedded. This criterion therefore stems directly from a belief that a unified body of scientific knowledge, in which the different parts of science are related to each other, are more consistent with each other, and illuminate each other, is in a very fundamental sense better—a greater intellectual achievement, more powerful, more beautiful—than an aggregate of scientific knowledge that is not so unified. If we believe this is what constitutes a more worthwhile science, our criterion of scientific merit is simply a reformulation of this postulate: pure scientific activities that unify scientific knowledge are better—that is, more valuable and therefore more deserving of support—than pure scientific activities that do not have such unifying results. The search for the unity of scientific knowledge is another version of the principle of Occam's razor: that science must be parsimonious, that it seeks to explain more with less. Thus, whereas utility is the value that guides the administration of applied science, and efficiency is the value that guides the administration of both pure and applied science, the unity of scientific knowledge should be the primary value that guides the administration of pure science.

To recognize the unity of scientific knowledge as a value in pure science is hardly new. What I have proposed is a fundamental distinction between truth and unity: truth and unity are both necessary objectives of scientific activity, but they apply to the two different aspects of science. Truth is the objective of scientific practice; unity of scientific knowledge should be the objective of the administration of pure science. Every pure scientific discovery or activity must satisfy a criterion of truth if it is to be admitted to the body of scientific knowledge, but, given the truthfulness of discovery, its value must be measured by the extent to which it contributes to the unity of the large corpus of scientific knowledge.

The sharp separation between truth and unity—the former being the underlying value of scientific practice, the latter of scientific administration—is perhaps too oversimplified. Eugene Wigner and Jacob Neufeld have pointed out that, although I claim truth to be the underlying value of scientific practice, I avoid giving a precise definition of truth. Is truth simply absence of internal inconsistency? Is it "correctness" in the sense that I can state my telephone number correctly? Or is truth itself to be judged in a larger context? Does the truth of a scientific proposition depend upon the degree to which that proposition explains other, neighboring scientific questions?

Clearly, scientific truth is all of these things, and, as Gerard Radnitzky says, much more—and I do not think that one definition of truth holds for every instance. In any event, I must concede that the truth of scientific theory—like quantum mechanics—is surely reinforced, if not proved, by the breadth of the phenomena that it explains—that is, by the unity it uncovers in the relevant sector of reality and the coherence it gives to the relevant science.

I cannot claim that the imperative of unity is the only explicit criterion used by the practicing basic scientist in charting his course. New questions arise in science from the falsification of older theories: a scientist chooses to examine a new question because he has encountered a "falsification."[9] Science is like a many-headed hydra, ever seeking to resolve falsifications. In short, every scientific question answered gives rise to new questions, and it is this ever-branching, self-generating activity that determines the structure of science.

Is there room for a commitment to unity in a scientific activity guided by its attempts to resolve falsifications? I believe there is, in the following sense: every question resolved gives rise to more than one unresolved question. Thus, falsifications present the scientist with a variety of choices for his next endeavor: he must still choose—either a simple problem or a more complex and fundamental one from among the plausible, and feasible, questions that he can attack. The criteria of choice that I have discussed, and in particular the criterion of unity, remain relevant. Falsifications, so to speak, provide the menu of problems; the criteria, and particularly the criterion of unity, underlie the scientists' choices among the problems.

Truth as the primary value for science has always been accepted without argument. To propose that truth ought to share this primary position with unity—the former in relation to scientific practice, the latter in relation to the administration of pure science—may appear to some to be unnecessary or simply odd. Yet it is not odd that the fundamental objectives of the two aspects of science are not necessarily the same. Perhaps the reason this distinction has not been taken as a matter of course in the past is because the profession of administrator of science, unlike that of scientist, is relatively new—too new to have commanded much attention among philosophers of science.

References

[1] J. B. Conant, *On Understanding Science* (Yale University, New Haven, 1947), p. 100.

[2] J. Bronowski, *Science and Human Values* (Harper Torchbooks, New York, 1959).

[3] A. Rapoport, *Science and the Goals of Man* (Harper, New York, 1950).

[4] A. Maslow, *The Psychology of Science* (Harper and Row, New York, 1966).

[5] M. Polanyi, *Minerva* **1**, 54–73 (1962).

[6] R. K. Merton, *The Sociology of Science* (University of Chicago Press, Chicago, 1973), especially Part 2.

[7] *Ibid.*, p. 204.

[8] A. M. Weinberg, *Minerva* **1**, 159–171 (Winter 1963); **3**, 3–14 (Autumn 1964); C. F. Carter **1**, 172–181; S. Toulmin **2**, 343–359 (Spring 1964); S. Rottenberg **5**, 30–38 (Autumn 1966). All the papers in the series were reprinted in *Criteria for Scientific Development, Public Policy, and National Goals: A Selection of Articles from Minerva*, E. Shils, ed. (MIT Press, Cambridge, MA, 1968).

[9] G. Radnitzky, *Revue internationale de philosophie*, CXXXI–CXXXII, 179–228 (1980).

The Philosophy and Practice of Scientific Administration

Scientific resources are allocated on both micro and macro scales. Micro choices are the innumerable allocations to individual researchers or small groups of investigators, largely conducted through peer-review procedures. Macro choices are the largest scientific and quasi-scientific decisions—SDI, the moonshot, the Super-Conducting Super-Collider (SSC), the Human Genome Project—that constitute much of national science policy. Has the philosophy of scientific administration (which I discussed in the two previous essays) influenced the way in which choices, either micro or macro, are made?

A. Micro Choices

At the micro level, I would say the influence, at least as judged by the National Science Foundation's practice, may have been considerable. I reproduce here the evaluation criteria taken from NSF's "Information for Reviewers," a document sent to all reviewers of proposals.

PROPOSAL EVALUATION CRITERIA

1. Research performance competence—Capability of the investigator(s), the technical soundness of the proposed approach, and the adequacy of the institutional resources available. Please include comments on the proposer's recent research performance.

Based in part on a lecture given at Cornell University, November 3, 1988.

2. Intrinsic merit of the research—Likelihood that the research will lead to new discoveries or fundamental advances within its field of science or engineering, or have substantial impact on progress in that field or in other scientific and engineering fields.

3. Utility or relevance of the research—Likelihood that the research can contribute to the achievement of a goal that is extrinsic or in addition to that of the research field itself, and thereby serve as the basis for new or improved technology or assist in the solution of societal problems.

4. Effect of the research on the infrastructure of science and engineering—Potential of the proposed research to contribute to better understanding or improvement of the quality, distribution, or effectiveness of the nation's scientific and engineering research, education, and human resources base.

Criteria 1, 2, and 3 constitute an integral set that should be applied in a balanced way to all research proposals in accordance with the objectives and content of each proposal. Criterion 1, research performance competence, is essential to the evaluation of the quality of every research proposal; all three aspects should be addressed. The relative weight given Criteria 2 and 3 depends on the nature of the proposed research; Criterion 2, intrinsic merit, is emphasized in the evaluation of basic research proposals, while Criterion 3, utility or relevance, is emphasized in the evaluation of applied research proposals. Criterion 4, effect on the infrastructure of science and engineering, permits the evaluation of research proposals in terms of their potential for improving the scientific and engineering enterprise and its educational activities in ways other than those encompassed by the first three criteria.

Criteria 1, 2, and 3 are more or less the criteria identified in the *Minerva* debate. Criterion 4, effect on the infrastructure of science and engineering, is rather different. It "permits the evaluation of research proposals in terms of their potential for improving the scientific and engineering enterprise and its educational activities in ways other than those encompassed by the first three criteria." The emphasis on education is particularly notable here since, in Frank Press's[1] recent ordering of scientific priorities, he has put activities that further scientific education at the very top. His rationale is, briefly, that *survival* of science requires a flow of new talented and informed young people; and surely survival must take precedence over every other consideration!

I have no data as to how seriously reviewers take these criteria in rating proposals. I suspect that the internal criteria—how competent the research has been in the past—probably figure largest in such judgments. This criterion is *a posteriori* whereas the others are *a priori*. It is much easier to judge a researcher's track record than to predict the outcome or significance of a proposed research. This is why I suspect that the *a posteriori* criterion may be taken the most seriously by reviewers, though I concede that this is largely a speculation. I hope NSF studies this matter so that we can base our estimate of the usefulness of criteria on more than anecdotes or speculation.

Macro Choices

During the 25 years that have passed since the debate on scientific choice began in *Minerva*, national science policy has been a recurrent concern of many governments and governmental advisory bodies. Thus, the United States National Academy of Sciences, in the late 1960s and through the 1970s, examined entire fields of science; among these were reports on physics (the first by a committee headed by George Pake[2] and the second by a committee headed by Allan Bromley[3]), on chemistry (under the chairmanship of Frank Westheimer[4]), and on biology (under the chairmanship of the late Philip Handler[5]). These studies reviewed the prospects and accomplishments of the fields under study, and, in at least some cases, attempted to suggest priorities among the various subfields. The theoretical debate on scientific choice figured fairly prominently in some of the studies, perhaps most notably in the Bromley study of physics, which appeared in 1973. Bromley and his colleagues not only promulgated a system of criteria by which each of eight subfields of physics and some sixty-nine "program elements" within the subfields were judged, they also assigned to each subfield and "program element" a numerical rating on each of thirteen criteria so that the "worth" of each subfield and "program element" was a "thirteen-component" vector. Bromley then suggested that priorities in the allocation of financial support ought to go to those fields whose "criterion vector" was longest. The criteria themselves were, with a few modifications, those discussed in the debate in *Minerva*. The major modification to the first criterion of choice set forth in *Minerva* arose from Bromley's recognition that certain very large and expensive pieces of equipment, like the National Accelerator Laboratory, required financial support in competition with subfields of physics, yet could hardly be judged by the same criterion. Instead, he constructed "structural" criteria that took account of the importance of these facilities to the pursuit of whole fields of scientific endeavor. I reproduce the ratings proposed by Bromley for the eight subfields of physics (Figure 1).

Although I have not followed the more recent attempts to allocate public funds among, and within, different scientific fields, the studies I am familiar with seem to invoke these or related criteria of choice—at least when they discuss the problem of arriving at a schedule of priorities, if not in the assignment of priorities to particular fields and projects. For example, the report of the National Academy of Sciences on research with transplutonium elements[6] repeatedly refers to the importance of research on transplutonium elements to the chemistry of all the transition elements, not simply to the chemistry of the transplutonics: "Chemistry is based on establishing interrelations in the behavior of the different elements. Each element, with its unique features, when its chemistry is compared with that of others, helps knit the fabric of theory that transforms a compilation of observations into a science."[7] In order to arrive at these judgments of the relevance of transplutonic chemistry to chemistry as a whole—its "scientific" merit, in my terminology—the panel sought advice from scientists in adjacent branches of chemistry and physics. Thus, insofar as the debate on scientific choice has provided a language, or an underlying structure, for the actual allocation of effort within science, I would say that the theoretical debate has been applied practically.

Nevertheless, one cannot claim too much for the criteria of choice as practical recipes for making choices in particular situations. Scientific choices at the highest level, involving many millions—even hundreds of millions—of dollars, often become political choices, determined not so much by abstract criteria of merit as by the interplay of competing political interests and objectives. This was seen particularly in the recognition by Bromley's panel that existing large facilities simply cannot be turned on and off. Moreover, the external criteria of merit can be applied to scientific activities much more easily after they have been completed than before. However, in a way, this is a deficiency of any criterion of choice, whatever the choice; we can decide on the wisdom of, say, a decision to go to war in Vietnam much better after the fact than before. Nevertheless, decisions must be made. That our criteria of choice for scientific administration are no better or no worse guides for actual administrative choices in science than are criteria of choice in any human endeavor must be regarded as a consequence of our inability to foresee the future, not necessarily a deficiency of the criteria themselves. Of course, the criteria were conceived mainly to help make the largest choices, the choices to do with big

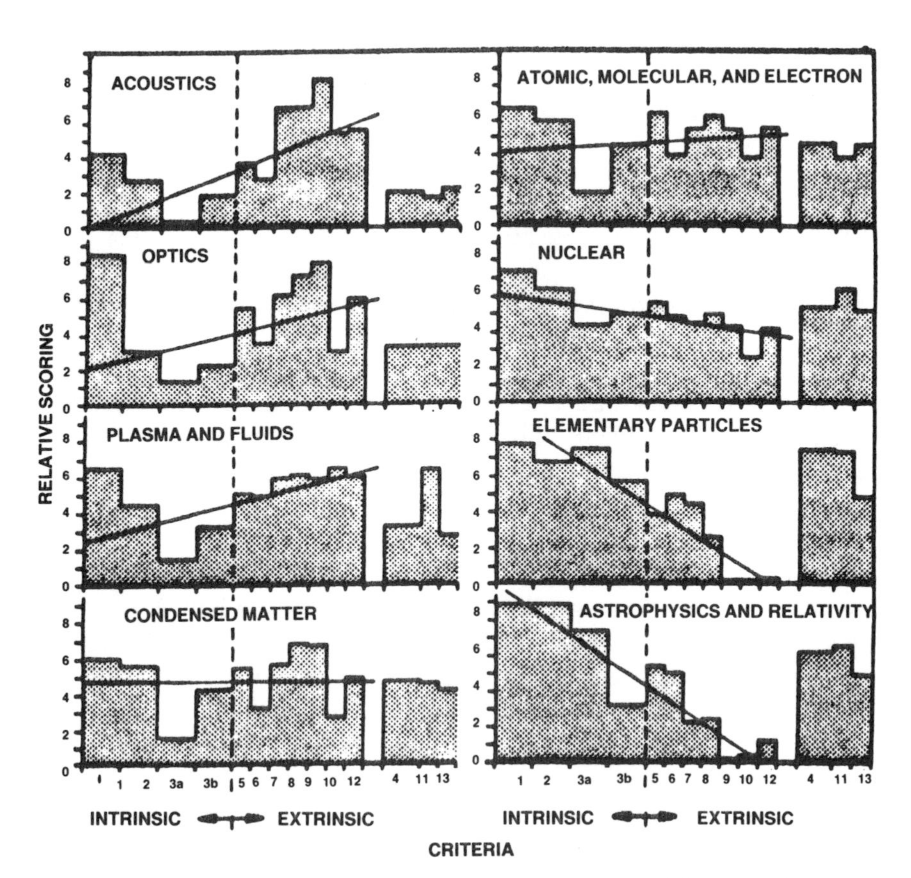

1 - RIPENESS FOR EXPLORATION
2 - SIGNIFICANCE OF QUESTIONS ADDRESSED
3a- POTENTIAL FOR DISCOVERY OF FUNDAMENTAL LAWS
3b- POTENTIAL FOR DISCOVERY OF GENERALIZATIONS OF BROAD SCIENTIFIC APPLICABILITY
4 - ATTRACTIVENESS TO MOST ABLE PHYSICISTS
5 - POTENTIAL CONTRIBUTIONS TO OTHER SCIENCES
6 - POTENTIAL STIMULATION OF OTHER AREAS OF SCIENCE
7 - POTENTIAL CONTRIBUTION TO ENGINEERING, MEDICINE, APPLIED SCIENCE
8 - POTENTIAL CONTRIBUTION TO TECHNOLOGY
9 - POTENTIAL FOR IMMEDIATE APPLICATIONS
10 - POTENTIAL CONTRIBUTIONS TO SOCIETAL GOALS
11 - CONTRIBUTION TO NATIONAL PRESTIGE AND INTERNATIONAL COOPERATION
12 - CONTRIBUTIONS TO NATIONAL DEFENSE
13 - CONTRIBUTION TO PUBLIC EDUCATION

NOTE: Histograms of the Survey Committee average jury ratings of the core physics subfields in terms of the intrinsic and extrinsic criteria developed in the Report. The straight lines superposed on the histograms are drawn simply to provide a characteristic signature for each subfield. It is interesting to note that these signatures divide naturally into three classes, with emphasis shifting from intrinsic to extrinsic areas as the subfield matures.

SOURCE: Physics Survey Committee, *Physics in Perspective* (Washington, DC: National Research Council/National Academy of Sciences, 1972).

FIGURE 1. *Ratings of eight subfields of physics: the Bromley Committee's proposals.*

TABLE 1. Big science spectaculars.

	Cost (billions of dollars)
Super-Conducting Super-Collider (SSC)	~6
Human Genome Project (HGP)	~3
Space Station (SS)	~30
Compact Ignition Tokamak Reactor (CITR)	~0.7
International Fusion Engineering Reactor (IFER)	~6
Ultra High Flux Nuclear Reactor (UHNR)	~0.5

science. If they also are relevant for choices in little science, where choices are less explicit, this is rather an incidental benefit of the criteria.

"Big-Ticket" Items

At the time I am writing (1988), there are at least a half-dozen "big-ticket" scientific proposals—Big Science spectaculars—that are receiving fairly serious attention, as well as some funding. I list them, together with a rough estimate of their cost, in Table 1.

Two of these, the SSC and the Human Genome Project, are purely basic research; the others are a mixture, with the UHNR being predominantly basic, the Space Station and the Thermonuclear Fusion projects being more or less applied (the Space Station may even have strong military implications). Can one attach priorities to these projects by invoking the criteria derived from the philosophic debate? For the applied projects, the criteria of usefulness are relevant, and surely they have been invoked by the proponents: if fusion energy remains a national goal, then CITR and (eventually) IFER are needed.

But for the two large basic projects—SSC and HGP—invoking the criteria seems to lead to vast disagreement. Thus, the high-energy physicists reject the notion that high-energy physics is rather isolated; ontological questions, by their nature, underlie all other questions and therefore almost by definition are related to more mundane questions. To this, solid-state physicists retort that they really have never invoked a finding in high-energy physics to further their branch of physics (for example, the Bardeen-Cooper-Schrieffer theory of superconductivity influenced the theoretical structure of high-energy physics, rather than the reverse).

Who is to decide between the high-energy and the solid-state physicists, or between the HGP and the rest of biology? The decision is ultimately, and necessarily, political. Just as President Kennedy's decision to go to the Moon was political and was taken without consultation of his science advisory committee—as was President Reagan's decision on SDI—so I recognize that the decisions on SSC and HGP will be essentially political decisions. The scientific community will not be unanimous either in support of or in opposition to these projects. Thus, how we finally choose on these spectaculars will depend strongly on how our politicians view our social priorities generally, as well as the state of our budget.

As for social priorities, I would argue that this list of science spectaculars ignores what I regard as the two most important long-term problems—AIDS and the greenhouse effect. As for AIDS, I am unclear as to whether spending gigabucks will help. But as for the greenhouse effect, there are very serious technologies, some very expensive, that could do much to forestall climatic catastrophe.

Three major thrusts are necessary: first, development and deployment of techniques for conserving energy; second, planting of trees to compensate, as best we can, for new fossil-fueled power plants; and third, development of cheap and transparently safe nuclear reactors that would allow the large-scale rebirth of nuclear power. I do not believe any one of these measures is sufficient; all are necessary. Thus I have proposed a worldwide yearly energy budget—by, say, 2040 A.D.—of 500 quads (quadrillion BTU's—half the usual scenario). Three hundred quads would be provided by 5000 large reactors; 150 quads would come from fossil fuel, of which perhaps 10 quads is tree-compensated (this requires planting 50 billion trees covering 4×10^5 square miles); and 50 quads would be provided by hydroelectric, solar, and geothermal. But even to contemplate this kind of rebirth of nuclear power is idle unless we restore the public's confidence in reactors. This would require *demonstration* of a new line of transparently safe reactors—reactors whose safety is passive and transparent. To demonstrate such a reactor—say, The Process Inherent Ultimately Safe (PIUS) or Modular High Temperature Gas-cooled Reactor (Mod-HTGR)—might take from 1 to 5 billion dollars—that is, somewhere between the HGP and the SSC in cost!

Where can we find all this additional money? After all, what we spend on basic science is quite arbitrary; indeed, as the economist Harry Johnson said almost 25 years ago, we always assume that what we are now spending is about right. Moreover, in respect to basic science's claim on the public budget, Johnson argued: "The argument that individuals with a talent for such research should be supported by society . . . differs little from the arguments formerly advanced in support of the rights of the owner of landed property to a leisure existence, and is accompanied by a similar assumption of superior social worth of the privileged individuals over common men. Again insistence on the obligation of society to support the pursuit of scientific knowledge for its own sake differs little from the earlier historical insistence on the obligation of society to support the pursuit of religious truth, an obligation recompensed by a similarly unspecified and problematical payoff in the distant future."[8]

Johnson's tongue was in his cheek when he said this. But he is right in implying that how much society spends on science depends on how much money is left after other needs of society are met.

I believe I see a possible source of the additional money, and this is as a result of the remarkable changes in the Soviet Union, exemplified by the speech, published in *Pravda*, of Vadim Medvedev,[9] new chief theoretician of the Soviet Communist Party, calling off the Cold War—indeed, rejecting the old idea of a death struggle between communism and capitalism. If these heady ideas prevail, then may we not in, say, the next decade see a serious reduction in the military budgets of the superpowers—say, from today's total of 600 gigadollars to, perhaps, 400 gigadollars? And should not some of this peace dividend be allocated to joint scientific projects?

A joint mission to Mars has been suggested. The International Engineering Fusion Reactor is now being planned at the International Atomic Energy Agency. The HGP already has international overtones. Why not consider doing SCSC as an international project? And, finally, the transparently safe reactor, motivated by concern over *global* warming—what is more appropriate than to make this a global project, especially since Mr. Gorbachev has already suggested something along this line.

I may have promised, at least by implication, more than I could deliver—that from the philosophic analysis would come clear directions for deciding between SSC and solid-state physics, between HGP and little biology. But, alas, this is not to be. Philosophy has never been that proficient in directing human affairs. Though the philosophic debate may give a language for, and even uplift, the political debate, the final decision will be, as such decisions have always been, political. Perhaps, over the coming years, we might avoid these choices by exploiting a new era of peace, where military budgets have been severely reduced and Big Science pursued internationally can expand our horizons as well as help to forestall environmental and public health catastrophes.

References

[1] F. Press, address delivered at One Hundred Twenty-Fifth Annual Meeting, U.S. National Academy of Sciences, Washington, DC, 1988.

[2] Physics Survey Committee, *Physics: Survey and Outlook* (National Research Council/National Academy of Sciences, Washington, DC, 1966).

[3] Physics Survey Committee, *Physics in Perspective* (National Research Council/National Academy of Sciences, Washington, DC, 1972).

[4] Chemistry Survey Committee, Committee on Science and Public Policy, *Chemistry: Opportunities and Needs* (National Research Council/National Academy of Sciences, Washington, DC, 1965).

[5] Committee on Research in the Life Sciences, Committee on Science and Public Policy, *The Life Sciences* (National Research Council/National Academy of Sciences, Washington, DC, 1970).

[6] Board on Chemical Sciences and Technology, *Challenges in Research with Transplutonium Elements* (National Academy of Sciences, Washington, DC, 1983).

[7] *Ibid.*, p. 62.

[8] H. G. Johnson, "Federal Support of Basic Research: Some Economic Issues," in *Basic Research & National Goals: A Report to the Committee on Science & Astronautics*, U.S. House of Representatives, by the National Academy of Sciences (U.S. Government Printing Office, Washington, DC, 1965).

[9] As quoted in the *The New York Times*, October 5, 1988:

He called for a new concept of socialism, borrowing political and economic ideas not only from other socialist countries, but even from the capitalist West.

"In working out the socialist perspective and formulating a modern concept of socialism, we cannot ignore the experience of mankind as a whole, including the non-socialist world."

On the question of the Soviet role in the world, Mr. Medvedev rejected the notion of a struggle to the death between Communism and capitalism.

"Present-day realities," he said, mean that "universal values" such as avoiding war and ecological catastrophe must outweigh the idea of a struggle between the classes.

"Peaceful coexistence, as we see it today, is a lengthy, long-term process whose historic limits are difficult to determine," he said.

Science, Government, and Information

Senator Hubert Humphrey was the unsuspecting godfather of the 1963 President's Science Advisory Committee (PSAC) report, *Science, Government, and Information: The Responsibilities of the Technical Community and the Government in the Transfer of Information* (SGI).[1] His Senate Subcommittee on Reorganization and International Organization had for several years been promoting an all-encompassing Department of Science. A major justification for such a department was the alleged disarray of the country's scientific information apparatus. Were all government-sponsored research placed in a single department, deficiencies in communication could be dealt with globally instead of piecemeal. This was the rationale offered for a Department of Science by the late senator and his enthusiastic staff.

The executive branch, under both President Eisenhower and President Kennedy, would have none of this. Putting all of government science under a single agency made no more sense than putting all accounting or, indeed, any other overhead activity under a single agency. Most government science is aimed at accomplishing a nonscientific mission, like stronger defense or better health. Thus, for example, to split military research and development, a means of achieving better defense, away from the Defense Department simply was wrong-headed. If scientific communication was deficient, the answer was to fix it, not start a new department! The President's Science Advisor, Jerome Wiesner, therefore responded to Senator Humphrey's chal-

Presented as the Joseph Leiter National Library of Medicine Lecture, April 7, 1988. Published in Bulletin of the Medical Library Association 77(1), 1–7 (1989). Reprinted by permission.

lenge by establishing a panel to study scientific communication, both inside and outside the government.

I was asked to chair the panel, not because I had strong credentials as an expert in scientific communication but, I suppose, because my credentials as Director of Oak Ridge National Laboratory (ORNL) were not clearly inferior to those of any other member of PSAC—that is, with the exception of William O. Baker. He had chaired PSAC's first panel on scientific information, which in 1958 had recommended the establishment of the Office of Scientific Information Services in the National Science Foundation. Dr. Baker was soon to become an emeritus member of PSAC and would then be technically ineligible to chair a panel. However, he was one of our most active and helpful panelists. The remaining panelists included two Nobel Prize winners, Eugene P. Wigner and Joshua Lederberg; a Krupp Energy Prize winner, Karl Cohen, manager of advanced engineering in the Atomic Power Experiment Department of General Electric Company; John W. Tukey, associate executive director of Bell Laboratories; James H. Crawford, Jr., a solid-state scientist from Oak Ridge who later chaired the physics department of the University of North Carolina; Louis P. Hammett of Columbia University, a past president of the American Chemical Society; Andrew Kalitinsky, a senior aeronautical engineer at General Dynamics; Gilbert W. King, an information scientist at ITEK; William T. Knox, an information scientist at Esso Research and Engineering Company; and Milton A. Lee, executive secretary of the Federation of American Societies for Experimental Biology. Francois Kertesz, a chemist who was also an information scientist at Oak Ridge, served as rapporteur and informal adviser. Jay Hilary Kelley, Office of Science and Technology, Executive Office of the President, was staff assistant.

Of the twelve panelists, three were physicists, four were chemists, two were mathematicians, two were biomedical scientists, and one was an engineer. Only three (King, Knox, and Lee) were professionally involved with handling of scientific information, though several others (Baker, Tukey, Crawford, and Lederberg) had given considerable thought to the management of scientific information. The rest of us were working scientist-engineers or research administrators; our approaches to issues were therefore hardly encumbered by prior knowledge.

Our panel met every month for about two years. At each meeting, we were briefed by professional scientific information experts both from government agencies, including the National Library of Medicine, and from professional societies. During the summer of 1962, I took leave from my job as director of ORNL and moved to Washington to draft our report. I would write every morning in my office at the Executive Office Building, I would then lunch with one or another of the many information specialists in the area, and then I would rewrite my morning's draft to incorporate what I had learned at lunch. By the end of the summer, my version of the report was finished. Our panel then met for another half year, arguing over each point in the original draft. The final draft was then reviewed by the entire PSAC and was issued in January of 1963.

Two points stand out particularly in my memory of the genesis of *Science, Government, and Information*: the great impact of Josh Lederberg's felicitous emendations and insights, and John Tukey's insistence that we "decrisify" the report. I had chosen the title "Science, Government, and the Information Crisis"; John argued that the issue would be with us forever and, therefore, "crisis" was an inappropriate word. I am still grateful to John for his measured wisdom in removing the original, exaggerated sense of crisis that permeated our first draft.

What Science, Government, and Information Said

The fifty-one-page report ranged widely: philosophic disquisition on the nature of the scientific enterprise; characterization of the "information problem"; advice to government agencies and to the technical community; and, less specifically, advice to librarians, documentalists, and individual scientists. Altogether the report contained eleven major recommendations. The tone of the report was exhortatory: its recommendations were couched in simple declarative sentences. As I reread it now, I blush at its pontifical tone—but, after all, it *was* a report carrying the imprimatur of the President of the United States!

Two main themes dominated: first, that handling of scientific information was, or ought to be, an integral part of science and, second, that retrieval of information was not the same as retrieval of documents. The first tenet was stated in the opening two paragraphs of *Science, Government, and Information*:

> Transfer of information is an inseparable part of research and development. All those concerned with research and development—individual scientists and engineers, industrial and academic research establishments, technical societies, Government agencies—must accept respon-

sibility for transfer of information in the same degree and spirit that they have accepted responsibility for research and development itself.

> The later steps of the information transfer process, such as retrieval, are strongly affected by the attitudes and practices of the originators of scientific information. The working scientist must therefore share many of the burdens that have been traditionally carried by the professional documentalist. The technical community must devote a larger share than heretofore of its time and resources to the discriminating management of the ever-increasing technical record. Doing less will lead to fragmented and ineffective science and technology.[2]

This statement, especially the last sentence, bespeaks a certain *hubris*: who can say whether science and technology are *too* fragmented and ineffective? Yet the specific recommendations—for example, to make the handling of information a part of an agency's research-and-development structure, not of the administrative structure—seem to me still to make good sense. Nor can I fault such exhortations to scientists as appeared later in the report, including "Write more clearly," "Write better abstracts and titles," and "Spend more time writing thoughtful review articles."

The second theme, that retrieval of documents was not the same as retrieval of information, owes its prominence in the report to Professor Y. S. Touloukian, head of the Thermophysical Properties Center at Purdue University. With the technology available in 1962, our panel could not conceive of supercomputers sensitively compacting literature with little or no human intervention. We therefore hit on the information center, manned by scientists as well as information specialists, as central to rationalization of the information system. Indeed, SGI's main legacy is still regarded as its visualization of the information center as a key, if not *the* key, to a future information system. The panel's position was stated as follows:

> The Panel sees the specialized information center as a major key in the rationalization of our information system. We believe the specialized information center should be primarily a technical institute rather than a technical library. It must be led by professional working scientists and engineers who maintain the closest contact with their technical professions and can make new syntheses with the data at their fingertips.[3]

We came to this position partly because of our puzzlement over the difference between documents and information (information centers, in contrast to libraries, purveyed information rather than documents) and partly because the chairman of the panel had firsthand experience with information centers at the ORNL. During the 1960s, the laboratory housed several information centers, most notably Kay Way's nuclear data center. This center provided annotated and standardized nuclear energy level schemes and cross sections for all known nuclides. Kay Way herself had started collecting the data at the Chicago wartime Atomic Energy ("Metallurgical") Laboratory and had continued the work at the National Research Council before moving to Oak Ridge. Other centers at Oak Ridge dealt with radiation shielding and reactor safety. In all cases, the staffs of the centers were professional scientists who spent at least half of their time digesting, evaluating, and summarizing the mountains of data that flowed from the many laboratories working in these fields. The report repeatedly stressed that the Information Center, like the centers at Oak Ridge, was part of a research laboratory, not part of a research library.

The report leaned over backward in its assessment of automation: "Mechanization can become important, but not all-important." Remember, this was before on-line terminals were common, and personal computers were not yet even a gleam in the eye of a very young Steve Jobs. Though we sensed that automation was coming, perhaps we downplayed its role because we were so anxious to elevate the position of the scientist in the information transfer chain.

Other recommendations dotted the pages of SGI. Citation indexing was a favorite, largely reflecting an enthusiasm shared by Josh Lederberg and John Tukey. Resident referees for government laboratories were proposed as a means of imposing higher standards on government contract reports. And we suggested that government departments that dominated, but did not monopolize, certain fields (as does NIH in biomedicine or AEC in high-energy physics) be designated "delegated agents" for controlling and organizing the flow of information in these fields. This suggested structure had a bit of the flavor of Senator Humphrey's Department of Science: information in each broad field of science would be the reponsibility of a delegated agency, even though research was performed in the field by several agencies.

Although these and other suggestions abound in SGI, the main focus of the report is its insistence that scientific information really is part of science and that the Information Center, staffed by working scientists, is a palliative, if not a panacea, for the information "crisis."

How SGI Was Received in 1963

Although SGI was aimed primarily at the research community and the agencies that supported research, the report had little impact on working scientists and engineers. It was not that the scientific community necessarily objected to or even lacked sympathy with the report's findings and recommendations; it was rather that the average scientist seemed to be unaware of the report's existence.

On the other hand, SGI created consternation, if not a sensation, among professional librarians. After all, the word "library" or "librarian" appeared not more than half a dozen times in the entire report, and generally in a vaguely scolding tone. Librarians were pictured as being somewhat anachronistic, as not keeping up with modern developments in automation or in new methods of bibliographic control. Perhaps most threatening to librarians was the call to the scientific community itself to take over functions that librarians had traditionally regarded as theirs. At least this is the impression one gets from the 1963 meeting of the Special Libraries Association (SLA). An ad hoc committee to study the report had been appointed to recommend actions SLA might take in response to SGI. Though several on the committee conceded that SGI had pointed to legitimate deficiencies in the way librarians dealt with the information onslaught, most of the committee regarded SGI's seeming dismissal of the role of librarians as reflecting an unjustified scientific bias on the part of SGI, and I was told at the time that one indignant librarian ceremonially burned a copy.

I illustrate the flavor of some librarians' reactions to the report with a quotation from Joan Morris's comments in the July/August 1963 issue of *Special Libraries*, the journal of the Special Libraries Association:

> According to the Weinberg Report, the present day librarian or information specialist will have to give up his work to a scientist who will evaluate the work of other scientists while the special librarian steps down to work of a clerical order.
>
> How such a report can be accepted so easily remains a matter of surprise. It either wholly disregards, or regards with insufficient attention, certain facts obvious to anyone who is working in the field of science—to wit, scientists capable in an area specialty have neither the time nor the interest to be information specialists for the nation.

> Area specialization among eminent scientists is now so deep that "hard" scientists have become myopic in outlook. . . . The truly impartial evaluators are the special librarians. Not only do they evaluate a scientist's offering on the basis of a wide acquaintance with the other offerings in the field, but they evaluate constantly on the basis of relevancy to the needs of their particular clientele.[4]

Perhaps twenty-five years after SGI, I ought to explain how this seeming antilibrarianism crept into SGI. When the President's Science Adviser announced to PSAC that a panel on scientific information was to be established, the late Professor Isadore I. Rabi, a distinguished member of PSAC, could hardly conceal his disinterest. "Humph—librarian's work" was his only comment. In the face of such indifference, our panel had to convince the scientific community that information was part of science, not simply "librarian's work." That there were legitimate deficiencies in librarianship twenty-five years ago could not be denied, but the report was not really aimed at the library community. It was aimed much more at the scientific community and at the agencies that supported science.

The appearance of SGI seemed also to exacerbate the rivalry between the community of documentalists and the community of librarians. On the whole, the report seemed to be more approving of documentalists, less approving of librarians, and I sensed at the time that the documentalists enjoyed what they interpreted as approval of their approach to information. Indeed, the information center (described in SGI as a research institute rather than a library) seemed close to the kind of organization documentalists considered to be their natural habitat.

The report became quite popular in the academic information community. After all, about a quarter of SGI was devoted to philosophic analyses of the role of information in the scientific enterprise, and some genuine issues in the philosophy of science were raised. For example, the report insisted that to understand the handling of scientific information, one must understand the very structure of science itself. Thus, some attention was devoted to how science divides into disciplines, how these disciplines relate to each other, and how they relate to the missions of the agencies that support them. The mission-discipline duality was a recurring theme in SGI, since findings in mission-oriented science were often of interest to discipline-oriented science; yet, because the two communities often interacted weakly, transfer of information between them could be laborious and inefficient. I have suspected that SGI's seeming popularity among academics was attributable to its raising such issues, which appeal to professors and students of information science.

Government agencies took SGI seriously. After all, it was a presidential document. Though no specific action was taken by the Federal Council of Science and Technology, SGI was taken as sanctioning information centers, and, in the years immediately following SGI, many new information centers sprang into being.

Science, Government, and Information in 1988

During the past twenty-five years, I have used information systems as an administrator, a government official, and a working researcher, but I have not been very close to the professional and governmental information-handling organizations that might have been influenced by SGI. Occasionally, as in preparing this paper, I have turned to my friends in the information community to brief me on the current situation, but by and large my impressions remain anecdotal and personal.

I have two main impressions: that information centers as conceived in SGI have not emerged as dominant elements of the information system and that SGI greatly underestimated the influence of the computer on today's information systems.

As for the specialized information center, some four hundred were in existence in the United States at the time of SGI. But this number had shrunk to only one hundred by 1980, when Carroll and Maskewitz reviewed specialized information centers in the *Annual Review of Information Science and Technology.*[5] The existence of such a review journal itself bespeaks the enormous development that has occurred in these years. According to Carroll and Maskewitz, SGI gave initial impetus toward development of information analysis centers, but the centers have not generally played the key role as synthesizers envisaged in SGI. Perhaps it was too much to try to impose on the scientific and technical community a new kind of social organization that did not evolve entirely naturally out of the needs perceived by practitioners themselves.

Although scientist-dominated specialized information-analysis centers had not emerged as the key element in the information system, on-line database services have expanded enormously. This growth is illustrated in Table 1, which appeared in the January 1988 issue of *Directory of OnLine Databases.*[6]

TABLE 1. Growth of On-Line Databases

Directory issue	Number of databases*	Number of database producers	Number of on-line services	Number of gateways
1979/80	400	221	59	
1980/81	600	340	93	
1981/82	965	512	170	
1982/83	1350	718	213	
1983/84	1878	927	272	
1984/85	2453	1189	362	
1986	2901	1379	454	35
1987	3369	1568	528	44
1988	3699	1685	555	59

*Includes distinctly named files within database families.

The databases in the table are not all scientific—indeed, a majority of these are legal and financial databases. Nevertheless, the tenfold increase in on-line services in less than ten years is astonishing. The directory attributes this growth to several technologies and information products, all of which have matured during the twenty-five years since SGI was published. These include databases (that is, collections of numeric or textual material processed for electronic publishing); timesharing computers that permit simultaneous access by many users; interactive computer programs; rapid-access storage devices; cheap computer terminals and microcomputers; and telecommunication networks.

To a degree, then, these on-line databases provide some of the services envisaged in SGI as being offered by information analysis centers. Generally, the databases do not provide analytical reviews, although in some cases they provide actual numeric data. For example, if I subscribe to the National Standard Reference Data System, I can call up on my desk computer the melting point and specific heat of uranium. Obviously, a great deal of analysis goes into the preparation of such databases. In some cases, the databases themselves are inputted by the working scientists. For example, the newly formed superconductivity information system, being developed by the Department of Energy's Office of Scientific and Technical Information, now has about two hundred subscribers, all of whom are active researchers in superconductivity. They provide abstracts, data, and short progress reports on their research to the central file. In return, they have full access to similar information from all the other subscribers. In a sense, then, the superconductivity data system provides relevant information to a dispersed company of researchers who in turn provide reviewed information to the system. Much of the function of an analysis center is therefore provided, although in a decentralized manner.

Another, largely unanticipated development since SGI has been the rise of serious scientific journalism. The scientific journalist, who is usually a highly qualified scientist turned journalist, now plays an important role as a reviewer and compacter of the scientific literature. *Nature* and *Science* today devote much more of their space to thoughtful reviews by professional scientific journalists and scientists than they did twenty-five years ago. On a more professional level, review journals are now much more common than they were twenty-five years ago. For example, *Annual Reviews*, which began originally to review the literature in biochemistry, now reviews twenty-seven separate fields of science and technology. And semi-popular scientific newspapers and journals now are common, whether as part of the *New York Times*, as journals like *Scientific American* or *Discovery*, or as newspapers for scientists, like Eugene Garfield's *The Scientist*.

Most of these secondary sources have been established by entrepreneurs who recognized an empty niche and proceeded to occupy it. This is true of the database services as well as the journalistic activities. These services more and more have had to meet the test of the marketplace.

Conclusion and Outlooks

As I contemplate the vast continent of scientific information systems, I realize that these systems, and their customers, are a diverse crowd. Schemes appropriate for some kinds of users are hardly relevant for other users.

In my view, the users fall, broadly, into two classes: first, those for whom time is of the essence, who operate necessarily in a state of perpetual "crisis" and who seem ready to pay money for their information services; and, second, those for whom deliberateness and studied response is of the essence, for whom haste makes waste, and who, in the customary scholarly tradition, expect information tools to be free services.

In the first category, I would put most engineering and operational scientists. Thus, a medical doctor who must diagnose a patient on the basis of various tests finds on-line medical diagnosis systems extremely helpful. For him time *is* of the essence, and on-line databases, particularly those embellished with modern artificial-intelligence methods, are obviously helpful.

Another example comes from my own field, nuclear energy. The accident at Three Mile Island (TMI-2) was a prime example of both information deficiency and information overload. The deficiency lay in the failure of the operators at TMI-2 to know that accidents almost identical to TMI-2, though less serious, had already happened at Davis-Besse and Rancho Seco. Had the TMI operators known of these, they surely would have diagnosed their problem before the core melted down. Information channels were overloaded; once the accident started, the control room was deluged with a bewildering avalanche of lights, bells, announcements, and data. What was desperately needed was a means of analyzing the raw data, extracting from it what was relevant, and presenting this *analyzed* data to the operator.

Since TMI-2, great progress has been made in providing operators with *relevant*, analyzed data. Indeed, just as medical doctors can now receive diagnoses on-line on the basis of the symptoms they input, so reactor operators can now receive advice on how to handle crises on the basis of information put into their artificial-intelligence control system. What we see here are examples of how prior intervention by sophisticated diagnosticians—either medical internists or reactor engineers—allow us to respond to crises intelligently. Indeed, the sophisticated application of artificial intelligence begins to approximate some of the functions originally conceived for the Information Analysis Center. I do not claim that many of the on-line databases are yet able to provide analyzed information that is useful in handling a time-constrained event; but I believe we are moving in that direction. In any event, the ability to call up data, even unanalyzed data, very quickly is generally helpful when a quick response is needed, whether in medicine or in reactor operations.

On the other hand, in research that requires the deepest kind of cogitation and that is not constrained by time, I am less than certain that having all data at one's immediate fingertips is always best. After all, the two most important discoveries of modern physics, relativity and quantum mechanics, were made long before on-line information systems were dreamed of. Could there be a kind of Gresham's Law—that, if information comes too easily, then researchers spend too much time absorbing the information and not enough time analyzing and contemplating its significance? Could our modern very quick response system be imposing on our scientific enterprise a sort of *journalistic* flavor that is antithetical to the deepest understanding of what is going on?

What I say must be heresy: that, if information systems provide data at *too* fast a rate, then the information process itself will tend to block out the processes of understanding and analyzing. Yet I should think this possibility must have occurred to many in the information community. Our plea for information analysis centers in a way was a response, twenty-five years ago, to our sense that something like this is a danger, or at least a consideration that we ought not to ignore.

Shortly after SGI appeared, I speculated on a possible role that information analysis centers might play in helping to codify and organize scientific knowledge and even to develop new, high-order scientific insights. My example was the development of the shell theory of nuclear structure. Ever since World War II, data on nuclear properties—cross sections, dipole moments, energy levels, and the like—have been accumulating at an astonishing rate. To find systematic regularities in this mass of data was not easy, yet, by the late 1950s, the nuclear information scientist Kay Way, with her encyclopedic command of the data, was able to discern periodicities in neutron cross sections as mass number increased. Helped by this inductive insight, Goeppert-Mayer and Jensen were able to devise the so-called shell picture of the nucleus. This model gave nuclear structure a coherence somewhat comparable to the coherence given to chemical structure by Mendeleef's periodic chart. I would say that, in this instance, the contribution of the information scientist, Kay Way, supported by her information center, was of signal importance.

Can we see an analogous development in biomedical sciences? I have in mind the project to map the human genome. As I understand the matter, the complete map would consist of some 10^9 base pairs. At the rate of 10^6 base pairs per day, the entire project will require some three years (and several billion dollars). Codification of the information and, perhaps more important, its analysis will obviously be as important and as challenging a part of the genome project as the identification of the bases in the first place. The biomedical community will be justified in asking "What do we do with the complete genetic map?" Just as the immense body of nuclear data was eventually reduced to some kind of scientific order by systematic organization of the data, I suppose it is fair to suggest, or perhaps to hope, that information scientists, perhaps even information centers, will play an analogous role in helping make sense of the billion-odd base pairs amassed by the genome project.

The genome project is hardly time constrained; nevertheless, its sheer magnitude and its enormous demands for data handling will surely place severe demands on those who will eventually try to make sense of the project. I should think that, along with newer, ever cleverer, and more sophisticated computers, the project will require that major data-analysis centers be devoted to it. And, insofar as the genome project is not time constrained, I would imagine the information-handling system will have less in common with on-line databases and more in common with the kind of information analysis centers originally contemplated in SGI. I realize these are speculations; yet it seems clear that the genome project will prove a major, possibly unprecedented, challenge to the National Library of Medicine.

Let me close on a different note. As I view SGI from the perspective of twenty-five years later, I realize that SGI's insistence on the specialized information center as being key to the resolution of the information problem was rather naive. History instead has shown that what prevails is not what a self-appointed committee of savants conceives but rather what the users, the working scientists, perceive as fulfilling their needs and, more and more, are willing to pay for. In the free-market competition between information tools, a variety of tools have emerged as dominant elements: on-line databases, review journals, new methods of electronic communication, libraries that take over some of the functions of information centers in providing information as well as documents, and, where the users feel the need, information centers themselves.

Nor is this likely to change. As I pointed out earlier, John Tukey insisted that we "de-crisify" SGI because handling information is not a problem that admits of a single neat solution, but it remains and always will remain a process that we must cope with. As Peter Medawar said in his 1969 address to the British Association for the Advancement of Science,

> We cannot point to a single definitive solution of any one of the problems that confront us—political, economic, social, or moral, that is, having to do with the conduct of life. We are still beginners and for that reason may hope to improve. . . . The great thing about the race was to be in it, a contestant in the attempt to make the world a better place.[7]

So it is with the information problem. Twenty-five years from now, on the fiftieth anniversary of SGI, I expect that the Leiter lecturer will return to these same themes and that the answers will be the same: that scientific information remains a serious issue, despite the existence of computers that make today's Crays look like toys; that the distinction between time-constrained and time-unconstrained tools will remain; and perhaps most important, that the National Library of Medicine will, as Peter Medawar implies, continue to be a major contestant in the race to maintain a viable scientific information system.

References

[1] U.S. President's Science Advisory Committee. Science, government, and information: the responsibilities of the technical community and the government in the transfer of information. Washington, D.C.: U.S. Government Printing Office, 1963.

[2] *Ibid.*, p. 1.

[3] *Ibid.*, p. x.

[4] Special librarians and the Weinberg Report. *Spec. Libr.* 1963 July/August; 54(6): 325–32.

[5] B. T. Carroll and B. F. Maskewitz, Ann. Rev. Inf. Sci. Technol. **15**, 147–89 (1980).

[6] Preface. Overall growth in the online database industry over the past eight years. Dir Online Databases 1988 Jan; 9(1): v.

[7] P. Medawar, Adv. Sci. **26** (127), 1–9 (1969).

Part III: Strategic Defense and Arms Control

Although I had never been directly involved in the design of nuclear weapons, my close contact with the original plutonium project sensitized me right from the beginning to the danger posed by the existence of nuclear weapons. I therefore participated in the founding of the original Federation of Atomic Scientists (later to become The Federation of American Scientists), and I was among the first of the young atomic scientists to testify before the U.S. Senate on the nuclear question. At the time, I was 30 years old. In my testimony before Senator Millard Tydings of Maryland, I insisted that the issue was war itself: that a nonnuclear war between two competent powers would inevitably end in a nuclear exchange if survival of one of the powers was at stake.

My rather unorthodox attitude toward nuclear arms—that defense was good, offense was bad—was much influenced by Eugene P. Wigner of Princeton. He spent a year at Oak Ridge National Laboratory in the mid-Sixties organizing a group to study civil defense. Even then, civil defense was regarded with great skepticism by most academic strategic analysts in the United States, although in Switzerland and in Sweden it has commanded serious attention.

My first article on defense systems, "Let Us Prepare for Peace," reflected views of the Oak Ridge Civil Defense study group. This was my first attempt to visualize a world in which mutually assured destruction was replaced by mutually assured survival. The occasion was the awarding of the Atoms-for-Peace Prize to B. Goldschmidt, H. D.

Smyth, and I. I. Rabi. My talk fell on very deaf ears indeed—a defensive world was no more popular among academics in 1967 than it was in 1983, when President Reagan visualized such a world in his famous Star Wars speech. In 1967, the nuclear community was still suffused with an unrealistic nuclear euphoria. This is evident in the visions of nuclear utopia with which I ended "Let Us Prepare for Peace."

President Reagan's Star Wars speech of 1983 evoked a responsive note from me, since he seemed to be supporting the position I had outlined in "Let Us Prepare for Peace." The idea for a defense-protected build-down (DPB), arose in the course of a discussion of the President's speech with Philip Morrison of MIT. Morrison had espoused a unilateral dismantling of U.S. offensive forces; I argued that such unilateralism was politically infeasible, unless each unilateral reduction in offense was compensated by a deployment of defense. This, in essence, was how the DPB idea was born. I learned later that similar ideas had been proposed by strategic analysts at Los Alamos. They apparently incurred the displeasure of the Air Force by pointing out that, if the U.S. is *not* intending to strike first, fewer MX missiles are required if the missiles are defended than if they are undefended.

The article "Stabilizing Star Wars," which I coauthored with my colleague Jack Barkenbus, is the first publication in which we outlined the idea of a defense-protected build-down. As described there, the DPB does not address the "ragged-retaliatory" issue—that is, even if the offense is reduced, the side deploying defense might be tempted to strike first since the defense is always better at handling a ragged retaliation than at blunting a first strike.

This point is central, and I deal with it in the next essay, "Deterrence, Defense, and the Sanctification of Hiroshima." I presented most of this essay at the Pugwash Conference in Austria in 1987. The ideas contained therein were worked out with Jack Barkenbus and appeared in a book we coedited, *Strategic Defenses and Arms Control* (Paragon House, 1987). That book and its companion, *Stability and Strategic Defense* (Paragon House and The Washington Institute, 1989), summarize a three-year study on strategic defense we conducted at the Institute for Energy Analysis under the sponsorship of The Washington Institute and the Lounsbery Foundation. In *Stability and Strategic Defense*, we collected half a dozen "game-theoretic" analyses that purported to show how a transition from today's naked offensive confrontation to defense dominance could be achieved without encountering crisis instability. The main conclusions of this game theorizing are summarized here.

The "Sanctification of Hiroshima," the last part of this essay, was presented at the International Conference on the Unity of the Sciences, Los Angeles, November 24–27, 1988.

With the ending of the Cold War, many of the heated arguments over strategic defense and deterrence may seem entirely out-of-date. Yet, as the partial success of the Patriot antimissiles in neutralizing Iraq's Scuds demonstrated, strategic defense against small attacks makes technical and political sense. This was really the central idea in Barkenbus's and my analysis, and I believe it makes sense even in President Bush's post–Cold-War "New World Order."

Let Us Prepare for Peace

The distinguished English economist Barbara Ward, in her book *Spaceship Earth*, suggests that the material abundance made possible by the new technologies will change qualitatively the relations among nations:

> May not the scientific and technological revolutions of our day produce a yet unguessed mutation in human attitudes? We have lived through the millennia on the basis of shortage. How will mankind react if relative plenty becomes the norm? In the past, conquest and imperialism, war and violence, have had their roots deep in the fact of absolute shortage. The desire to take your neighbor's land, to lay hold of his resources, to overcome your inadequacies by making his life more inadequate still—have not these been, again and again, the bitter causes of aggression? And insofar as nations recognize the . . . need for "living space," they almost instinctively choose for leaders men who articulate these violent needs and envies. Prosperous people very rarely choose lunatics for rulers.

Most nuclear scientists, by and large, believe in Ward's optimism; they are sustained in their endeavors by her vision of an abundant peaceful world. The developments in nuclear energy, notably those of the past few years, seem to be fulfilling this vision. Nuclear power, in 1952, was written off by a distinguished scientist with the prediction that, in the 1960s, the effort toward developing nuclear power would be abandoned (*Report by the Joint Committee on Atomic Energy, Atomic Power and Private Enterprise*, page 330, U.S. Government

Presented at the Seventh Atoms-for-Peace Award Ceremony, Rockefeller University, November 14, 1967. Published in Bulletin of the Atomic Scientists **24**, 7 (1968). Reprinted by permission.

Printing Office, Washington, D.C. [December 1952]). Today—in Canada, in France, in the United States, and in the United Kingdom—nuclear power is a competitive source of energy. For example, in the United States we now have on order, in operation, or under construction close to 60 million kilowatts of nuclear power. This represents almost a quarter of all central electric power capacity in the United States, and this conversion to the atom shows no sign of abating.

But we are only at the beginning; we have still not fully exploited either the ubiquity or the intrinsic cheapness of nuclear energy. Because nuclear energy is not tied to cheap indigenous fossil fuels or to swiftly flowing rivers, it can be placed wherever energy is needed. Thus, for example, we can visualize large nuclear plants springing up in arid coastal deserts to energize large desalting plants. The technology for large-scale desalting is here, and the costs are reasonable even when the evaporators are energized by conventional reactors. The by-product electricity can be used to manufacture fertilizer and reduce metal ores and to light cities. Altogether we see this ubiquity and mobility of nuclear energy making possible a kind of nuclear-powered agro-industrial complex that could give practical embodiment to Ward's notion of a material autarky throughout the world. This general line of thinking underlies the proposal now before the United States Senate to deploy such nuclear complexes in the Middle East and, in effect, provide a new framework of physical resources in which to seek a resolution for that region's desperate animosities.

Thus, eventually, when the advanced breeders are developed, many if not most of our material wants will be satisfied by energy from fission. And, insofar as low-grade thorium and uranium ores are available everywhere, each region of the globe and each country will have its sources of very cheap, abundant energy. This energy will be converted into the water and the fertilizer and the food and the metals on which civilization depends. The world should become immeasurably richer than it is today. Neighbors would no longer scramble for each other's green pastures, and there would be a general easing of tension once want is eliminated.

The Preconditions for Peace

This vision of a *Pax Atomica*, of a world in which tensions have relaxed because scarcities of raw materials are no longer rational bases for conflict, is a golden vision, one to which all of us in the nuclear business are dedicated. And yet it is an incomplete picture of the peaceful world of the future. It neglects those sources of strife that are not rooted in geographic inequities or disparities in natural endowments. There remains the strife that comes from ideological conflict and conflicts of interest, the strife that comes from the all but universal human ambition for influence or power. Our atomic-powered Utopia needs more than material well-being, important as that may be, to stabilize the *Pax Atomica* and to prevent war.

But, even more, this vision ignores the present nuclear confrontation between the superpowers. It has been customary to look to the hydrogen bomb and mutual deterrence as the means to prevent war, to curb the largely emotional drives that impel men in power to seek to maintain their positions or to extend their influence. And, a little surprisingly, the balance of terror has worked—not perfectly, but still tolerably well. We have had wars since the atomic bomb was used in Hiroshima; but we have avoided all-out world war and with it the thermonuclear holocaust.

Yet most of us are acutely uncomfortable with this balance of terror, wherein the two superpowers hold as hostages 100 million of each other's citizens. It is unprecedented in world history that the citizens of the strongest powers in the world can no longer be guaranteed by their state some measure of personal security, except to the extent that the balance of terror dissuades the other side from striking. Somehow, one is appalled by the possible fragility of this metastable balance.

It is largely on this account, this nervousness about the stability of the balance of terror, that the world has wrestled mightily with arms control and disarmament. Moreover, the nuclear world of plenty is inconsistent with a world in which ever-increasing portions of the gross national product might go into maintaining the deterrent. It seems apparent that we must ultimately disarm; but how can we both disarm and maintain the deterrent? How can we get from here—a world filled with mutual apprehension, with ICBMs, with megaton warheads—to a world based on energy self-sufficiency, mutual respect, and peace? How can we, as Amrom Katz of the Rand Corporation says, make the world safe for disarmament?

Hope in Failure

I believe, paradoxically, that a way may have been opened by the failure of the negotiations over deployment of antiballistic missiles. The deployment of ABMs on both sides has been deplored as a new step in an unending arms spiral that eventually will consume everything, including our vision of abundance. But suppose ABMs and other defensive measures turn out to be effective and, at the same time, there is no escalation of offense in unending spiral. The knife-edge of delicately balanced terror would then be blunted. Perhaps then, as D. G. Brennan of the Hudson Institute has stressed persuasively (*Bulletin Of The Atomic Scientists*, June 1967), we should not be so disturbed if the threat of ultimate, absolute, and total mutual destruction is not forever to be the basis for our world order.

If there is even a remote possibility of achieving effective defense and at the same time limiting offense, should we not examine, very much more carefully than we have, the possibilities of an essentially defensive posture? Granted that active defense systems today are not perfect, they nevertheless seem to be much more effective than they were thought to be five years ago. And, by virtue of the development of the admittedly imperfect and light antiballistic missile system, we have already achieved a kind of *de facto* disarmament. Because space and weight in offensive rockets must be allocated to penetration aids, the total number of megatons each side can throw at the other ought to be reduced by the antiballistic missile. In this sense, the ABM has caused a kind of arms limitation, one of the few real arms limitations that we have achieved.

Moreover, passive defense, a subject about which we hear very little, may be much less impractical than is commonly believed to be the case. We at the Oak Ridge National Laboratory have been studying the question of civil defense for the past three years under the guidance of James Bresee and Eugene Wigner. The result of our studies suggests that underground, interconnected tunnels used as shelters could significantly reduce the casualties caused by thermonuclear weapons. In this connection, I remind you that at least one distinguished city planner, Constantinos A. Doxiadis, holds that the megacity of the future can survive only if it puts its transportation (including automobiles) and utilities underground. The megacity will therefore, according to Doxiadis, be honeycombed with tunnels. Such tunnels would be the main elements of a passive defense system: that they might come rather as a matter of course as the city develops should not make them less practical for dual use as shelters.

But we are told all this is transitory: antiballistic missiles and civil defense will be followed by more ICBMs, which will be followed by more ABMs and more civil defense in an unending spiral. We shall go from 3,000 megatons to 30,000 megatons to 3,000,000 megatons—where does the crazy spiral stop? It is here that Brennan has injected a new and elegant idea into the discussion: Should not the world, in negotiating the next perilous stage of arms control, focus primarily on limiting offensive weapons and, at the same time, encourage defensive systems? All the predictions about deployment of antiballistic missiles and civil defense leading to unending escalation assume that offense will escalate indefinitely. But if the world agreed to, and enforced, a limit on the number of ICBMs, we would stop the spiral of escalation. Such limitation on primary instruments of offense are not unprecedented. In the post-World War I era, capital ships of the three great naval powers were limited. Moreover, if defensive systems continue to improve, the capacity of the world to destroy its people and its lands will gradually deteriorate; and the number of hostages held on each side will be reduced, though certainly never to zero, so that nuclear war, even in a defensively oriented world, could never be regarded as a rational instrument of policy.

The difficulties of such a posture and such an agreement—to limit offense but leave defense unlimited—are formidable. Can one police a freeze on offense unless secrecy is relaxed? Will such an arrangement withstand pressure for abrogation by those who underestimate their own offense and overestimate the opposition's defense? Will strong defense tend to make each side more aggressive in the conduct of its foreign policy? These are imponderables, but one must remember that the present balance of terror is not a lovely thing to contemplate, nor is it a perfect antidote against thermonuclear war. L. B. Sohn of Harvard suggested that an existent posture need be only 50 percent foolproof, while a newly proposed posture must be 98 percent effective. If we addressed as much time and energy to developing the details of a defensive posture in arms control as we have devoted either to developing offensive armaments or to formulating present arms-control doctrines, is it not at least possible that we would be able to work out credible answers to many of the difficulties we now see in limiting offensive weapons?

There are two overriding reasons why we must eventually come to some such position. The first is that, much as some deplore it, both of the superpowers have decided to deploy antiballistic missiles. We are in grave danger of an unending arms spiral unless we enter into agreements to chop off the spiral at the top. This implies some limitation—possibly tacit, but preferably explicit—on, say, the total number of offensive missiles or on the total expenditure for offensive missiles.

There is another reason that seems to me even more compelling. Can we ever hope to achieve real arms control or disarmament from the present position of overwhelming offensive power and almost nonexistent defense? Does anyone really believe, in the kind of hard, untrusting world we live in and shall have to live in during the next several decades, that either side will agree to a disarmed world unless it feels secure in its defensive systems? Can we realistically contemplate disarmament with the possibility of clandestine sequestering of a few missiles, or without being reasonably certain that our defenses can handle a sudden attack from such missiles?

What Posture for Peace?

In the main, our military technology has emphasized offense rather than defense, and our arms-limitation technology has emphasized defense rather than offense, especially in the most recent discussions of the antiballistic missile. I submit that both postures may have been in error and that the cause of peace will be better served by developing ways to strengthen defense and to limit offense.

I would therefore urge that the military communities of the world prepare for peace by developing defensive systems, rather than continuing to exert themselves primarily in improving offensive systems. And I would urge that the arms-limitation communities of the world prepare for peace by developing doctrines for limiting offense and techniques for enforcing such limitations, rather than continuing to exert themselves primarily in limiting defensive systems.

It seems that herein we may find the missing elements in the world described by Barbara Ward. We shall have our cheap nuclear power and our agro-industrial complexes and our energy autarkies. But we shall need something other than the balance of terror to keep the peace in the long run. World government, or general and complete disarmament—these are mere words unless we see credible ways to go there from here. The energy-rich world, even with most of its material wants provided for, will still be a world of nation-states, each with its own imperatives and traditions and glorious history—and its habits of violence. And this world will for a long time have its military establishment. Does not common sense dictate that a world whose military is preoccupied with defense rather than with offense is more rational than the bizarre and precarious world we now have, a world we would be contemplating with horror if we were not so tired of its grim countenance?

Stabilizing Star Wars

WITH JACK N. BARKENBUS

For more than three decades, the arms race has spiraled upward because neither the United States nor the Soviet Union will accept anything less than strategic parity with its rival. Clearly, traditional arms-control efforts show little promise of reducing the nuclear stockpiles, even though deterrence with much smaller arsenals is desirable and theoretically possible. Suggestions to speed up the arms-control process with unilateral actions fail to take into account the political realities of today's world. Given prevalent strategic doctrines and the current climate of tension and distrust, no leader will order a serious reduction in nuclear weapons without an equivalent and verifiable reduction by the other side.

But, by exploiting an unfolding change in U.S. doctrine, unilateral measures with a real chance of prompting a positive Soviet response might be politically feasible. Unilateralism can work, if the United States dismantles some offensive weapons at the same time it deploys ballistic-missile defense (BMD). The dangerously high number of offensive strategic missiles in the world can be reduced if Washington embarks on such a "defense-protected build-down (DPB)."

Although most analysts agree that, in theory, a defense-oriented

---

This essay, which is co-authored by Dr. Jack Barkenbus, first appeared in *Foreign Policy*, No. 54, Spring 1984, 164–170 (1984). Reprinted by permission.

world is safer than, and preferable to, an offense-oriented world—that mutual assured survival is better than mutual assured destruction (MAD)—defensive weapons systems are widely opposed for three reasons. First, many arms-control experts fear that deployment of these weapons will touch off yet another expensive round of U.S.–Soviet technological competition, leading eventually to new offensive weapons systems that can overwhelm the defense. Second, even if this renewed competition for offensive superiority does not occur, it is argued, any defense that is not totally effective against awesomely destructive nuclear weapons is not worth deploying. Finally, the transition from an offense-oriented world to a defense-oriented one is generally seen as fraught with risks. At worst, the deployment of defensive weapons by the United States might compel the Soviet Union to deploy its own defensive system and expand its offensive forces—or even to attack the United States before its own forces could be neutralized.

An orderly transition to a defense-oriented world, however, can be achieved by combining deployment of defensive weapons with a concomitant and compensating reduction of offensive weapons. By destroying a certain percentage of its missiles in the process of deploying defensive weapons systems, Washington would clearly signal Moscow that the United States was not seeking a strategic advantage but simply exchanging one kind of parity for another.

A New International Norm

The DPB scheme is quite different from the vision President Ronald Reagan conjured up in his Star Wars speech of March 23, 1983, of an America—and ultimately a world—free of the threat of nuclear destruction. That vision is an illusion. Science and technology cannot rid the world of nuclear weapons. Even if a perfect missile defense were possible, other delivery systems for nuclear weapons would still endanger millions of Americans. The real issue, therefore, is not whether the world can return to an era free from the possibility of a nuclear holocaust, but whether political leaders can significantly reduce the level of destruction that can be wreaked on the Earth—that is, reduce the force level at which MAD operates.

The Star Wars speech called upon the scientific community to devise a network of exotic space- and land-based systems capable of intercepting and destroying Soviet strategic ballistic missiles during various stages of their flights. No one not privy to the classified scientific literature on these systems can definitively judge the feasibility of these programs. It appears abundantly clear, however, that BMD systems using such devices as lasers and particle beams would be extremely expensive, vulnerable to much less expensive Soviet countermeasures, and deployable only in the twenty-first century.

DPB, however, does not depend on the success of this costly technological gamble. Instead, it is based on the development of BMD systems using interceptor missiles, which has progressed steadily over the last decade. Short-range interceptors capable of defending offensive missile sites—not population centers—by attacking Soviet warheads as they reenter the atmosphere can probably be deployed in the 1980s. The uncertainties are great, but at this point the best evidence suggests that such systems could destroy between two and eight reentry vehicles for every one that gets through.[1]

Although many strategic defense enthusiasts bemoan the fact that current BMD systems could only provide marginal improvements in U.S. defenses, this characteristic is actually a major advantage of DPB. Its incomplete effectiveness requires policy makers to move incrementally, thereby preventing a sudden destabilization of the strategic balance. The importance of DPB in the beginning resides not in the level of protection provided but in its ability both to initiate a winding-down process and to establish a new international norm of behavior based on a reduction of offensive arms rather than an expansion. DPB should therefore be limited initially to the construction of silo-defense systems and concomitant reductions in very accurate U.S. land-based intercontinental ballistic missiles (ICBMs) that, because they threaten Soviet forces, raise the specter of a first strike.

The call for simultaneously deploying defensive systems and freezing the levels of offensive weapons—for creating a lightly armed, heavily defended world—is not new, since strategic planners have long recognized that the unlimited use of offensive weapons can invariably overcome defensive systems. What is new about the DPB proposal is the symmetrical reduction of offensive forces based upon credible estimates of the defense's effectiveness.

Assume that, initially, without defensive systems, both the United States and the Soviet Union deploy an equal number of offensive weapons. A BMD system deployed by the United States would reduce the number of Soviet weapons able to reach their targets, upsetting the strategic balance. Under DPB, the United States would restore this balance by reducing its stockpile of offensive weapons enough so that the two sides would be again evenly matched in warheads that can hit their targets.

Here is a simplified example of how DPB might work. Assume that the United States and the Soviet Union have achieved parity with 1,000 weapons each. A U.S. BMD system capable of destroying 10 percent of these warheads in an all-out Soviet attack would leave Moscow with only 900 deliverable weapons. This situation would permit Washington to dismantle 100 of its warheads and still maintain the offensive balance. If the Soviets followed suit by deploying a BMD they believed would be twice as effective—which would have double the kill probability of its U.S. counterpart—Moscow would have to reduce its nuclear arsenal to 800 warheads. These measures would leave each side with 720 deliverable warheads in each other's estimation, since the United States would consider itself able to destroy 10 percent of the Soviets' 800-warhead arsenal, and the Soviet Union would consider itself able to destroy 20 percent of America's 900-warhead arsenal.

The reduction in American offensive weapons would give the Soviet Union a powerful incentive not to increase its offensive forces or deploy a BMD without scrapping any missiles. If Moscow tried to take advantage of the situation, the United States would be forced to restore some or all of its destroyed offensive weapons. Hence, the traditional argument against defensive systems—that they would cause the offensive arms race to escalate—would no longer hold. Under DPB, the pressure to catch up would be eliminated because the deployment of defensive weapons and the concomitant destruction of a certain number of offensive weapons would maintain the strategic nuclear balance.

The difficulty in estimating the effectiveness of a defensive system is a serious shortcoming of the DPB strategy. The United States will prudently tend to underestimate the effectiveness of the American defensive system and thus dismantle as few of its offensive weapons as possible. The equally prudent Soviet Union, in responding to the deployment of a U.S. BMD system, could be expected to overestimate the effectiveness of American defenses.

To overcome this difficulty, DPB should be implemented in very small stages. If the initially deployed American BMD system were judged capable of neutralizing one-tenth of the Soviet force, the United States would have to reduce its own warhead arsenal by only 10 percent. Thus, even large errors in the kill-probability calculation would affect the weight of a delivered Soviet attack only marginally.

Therefore, the United States could afford to implement unilaterally a DPB policy and assess the Soviet response. There are risks and expenses in dismantling a small fraction of U.S. offensive arms should the Soviets choose to bolster their offensive capabilities in response to DPB. But the risks would at worst be temporary. The United States would not continue to build down if the Soviets reacted negatively.

A Defense-Oriented World

Although DPB should be implemented incrementally, these increments could add up, over time, to substantial protection. Each reduction in the adversary's offensive warheads would invite a corresponding reduction in the other side's offense without requiring a formal treaty. The scheme should lead to a downward rather than an upward spiral, provided each side takes its own defensive potential seriously and actually reduces offensive forces as defensive systems are put into place. The actual attrition of the offensive force could produce a world that, while still dependent on MAD to keep each side honest, would be threatened by far less astounding force levels. Yet there is no magic force level above which MAD holds and below which it loses its potency. Even one 100-kiloton bomb aimed at each of the 10 largest cities in the United States and the Soviet Union would be a very formidable deterrent. The attractiveness of disarming through defense is its potential to shrink force levels in terms of both weapons launched and weapons delivered. And if one views the primary purpose of the defensive weapons to be the creation of a situation that allows political leaders to reduce their own arsenals of offensive weapons rather than to neutralize an attack totally, the demands on the system are correspondingly reduced.

Implementing DPB would require withdrawing from the 1972 U.S.–Soviet antiballistic missile (ABM) treaty—a step that should be taken only with the greatest forethought. Abrogating this treaty would heighten superpower tensions if it were done precipitously and without regard for creating a better substitute. But if the ABM treaty were amended to allow for gradual, explicitly defined phasing in of BMD systems together with DPB, the agreement would stabilize the build-down process. Moreover, if the administration's strategic-defense policies proceed much further, they may violate the ABM treaty anyway. DPB could help assure that the treaty's termination has constructive, not destructive, consequences. Indeed, the expensive ABM research Reagan favors can be justified only if the eventual deployment would diminish, not exacerbate, superpower tensions and slow the offensive arms race.

Recent frustrations in arms-control negotiations emphasize the advantage of a defense-oriented strategy. Arms controllers have always realized that a two-party MAD confrontation is theoretically stable, for parity is achieved when both sides have the same number of missiles. But a three-party MAD confrontation is theoretically unstable, since each party will seek as many offensive weapons as its two adversaries combined. A lasting parity is possible only if no parties have any missiles or if each party has an infinite number of missiles. This problem beset the now-suspended intermediate nuclear force negotiations; the Soviets insisted on deploying enough medium-range missiles to counter French and British as well as U.S. forces, while the United States demanded parity based on the superpowers' medium-range arsenals alone. By contrast, a defense-oriented world would be less prone to this instability, since defense without offense is not threatening. And if eventually both superpowers were armed to the teeth with defensive systems but all nuclear powers had only a few offensive weapons, nuclear war would not necessarily lead to nuclear annihilation.

Ballistic missile defense has until now been attacked as a costly, unworkable, and destabilizing pipe dream, advocated only by recalcitrant hawks intent on building a Fortress America. DPB provides a way of using today's imperfect defense technology to reduce the superpowers' arsenals and the extent of the catastrophe that would result should deterrence fail.

The world will never again be free of the weapons of mass destruction. Disarmament alone can never be a practicable mechanism to achieve this world, a world that is not hostage to these devices. But by committing itself to both offensive disarmament and the deployment of defensive weapons, the United States can begin to put an end to the senseless, perilous, and seemingly interminable arms race.

References

[1]A. B. Carter, lecture delivered at the American Physical Society meeting, 1983.

Deterrence, Defense, and the Sanctification of Hiroshima

Ever since the Soviet Union acquired the capability of attacking cities with thermonuclear weapons, fear of retaliation has deterred nuclear war. Despite targeting doctrines that abjure attacks on cities ("countervalue"), the collateral damage from an attack on silos ("counterforce") all but eliminates the difference between the damage caused by countervalue and counterforce attacks. These last forty-odd years are therefore properly regarded as the MAD era—the era of Mutually Assured Destruction: Peace has been maintained by fear of assured retaliatory destruction.

No one knows how long the MAD era will last. Although the superpowers have avoided conflict, many have doubted both the morality of a posture that depends upon an implied threat to kill—even inadvertently—millions of innocents and the robustness of the standoff in the event of extreme international tension. Politicians as well as analysts have therefore tried to visualize an alternative that was as effective as MAD in keeping the peace, yet was not encumbered by the profound moral deficiencies of MAD.

The most important attempt to move away from MAD was initiated by President Reagan's March 23, 1983, "Star Wars" speech, which launched the Strategic Defense Initiative (SDI). The president called

---

Based on a paper presented at the 17th International Conference on the Unity of the Sciences, November, 1988, Los Angeles, California. See also *The World and I*, June 1989, pp. 467–491. I wish to thank Dr. J. Barkenbus for his many contributions to this discussion.

for transforming the MAD era into the MAS era, the era of Mutually Assured Survival. This was to be done by development and deployment of defensive systems, mostly space-based, that would render nuclear weapons "impotent and obsolete." Peace between the superpowers would then depend not on fear of retaliation but upon denial of aggressive intent.

Many analysts as well as religious leaders concede that MAS, if it works in keeping the peace, is morally superior to MAD. On the other hand, most, though by no means all, analysts doubt that MAS is feasible technically; or, if it is, that a transition from MAD to MAS can be managed without encountering insuperable instabilities.

The purposes of this paper are threefold. The first task will be to describe the MAD and MAS eras and to compare the moral sanctions for each. This will require an elementary exposition of some principles of moral philosophy. The second purpose will be to summarize several recent studies that purport to show how the transition from MAD to MAS can be made without encountering either crisis or arms-control instabilities.

But this is not enough. Even though the world may shift from MAD to MAS, can the MAS era last forever? The technologies of nuclear fission and fusion will always be part of our heritage. Can we live forever with the awful knowledge of how to destroy ourselves without actually doing so? I shall therefore speculate about what I designate as the era of nuclear renunciation: the era in which nation-states have become peaceful and in which taboos have been erected around nuclear explosives—taboos powerful enough to exorcise the threat of nuclear annihilation completely and forever.

The Moral Sanctions of MAD and MAS*

Joseph Nye, writing in *Nuclear Ethics,* explains the ethical principles that underlie the debate on nuclear deterrence. What follows is in large part a paraphrase of his excellent disquisition.[1]

Two major and contrasting principles have dominated Western ethical thinking for at least the past century. The first, which goes by the formidable name of deontology, finds its roots in the ethics of Imman-

*The sections entitled "The Moral Sanctions of MAD and MAS" and "The Morality of Offensive Deterrence, Disarmament and Defense" are largely taken from A. M. Weinberg and J. Barkenbus, "Morality, Arms Control and Strategic Defenses," in *Arms Control After SALT,* edited by Steve Cimbala (1988) (SR Books, Wilmington, Del.).

uel Kant. This principle holds that an act is moral, right, and acceptable if it conforms to certain prior rules of behavior, quite apart from the consequences that result from the act. Thus, since one of God's commandments is "Thou shalt not kill," a thoroughgoing deontologist would not kill an assailant even if, by not killing the assailant, he would risk losing his own life. A thoroughgoing pacifist would be an example of a complete deontologist. In general, deontology is concerned with means, not ends: What is moral is what utilizes moral means.

The other, contrasting, view is called consequentialism. It traces its origin to the utilitarian philosophers Jeremy Bentham and John Stuart Mill. According to this view, an act is moral if the consequences of that act are moral, quite apart from whether the act itself is moral as judged by other standards of behavior. Thus, a thoroughgoing consequentialist would condone killing in self-defense, even though this act violates one of God's commandments. In general, consequentialism is concerned with ends: What is moral is what results in moral consequences.

These examples overdraw the distinction between deontological and consequentialist reasoning. In practice, the deontologist always considers consequences, just as the consequentialist considers rules. But, by and large, the arguments over the morality of nuclear deterrence have been either predominantly deontological or predominantly consequentialist.

Nye introduces a third moral dimension—motives. For him, an act that obeys prior rules of behavior and leads to morally acceptable consequences may still be morally unacceptable if it is basely motivated. Of course, the judgment as to whether a motive is base itself must depend upon the prior commitments and values of the judge. Many Americans still regard our involvement in Vietnam to have been motivated by a morally unassailable principle—the defense of freedom; others might claim that the distinction between Diem's regime and the communists was not as morally unambiguous as apologists for U.S. intervention claimed it to be, and our motives were therefore suspect.

The Morality of Offensive Deterrence, Disarmament, and Defense

Despite the complexities and endless elaborations that accompany the application of the foregoing distinctions to security policies, the case for the morality of alternative security positions can be stated forthrightly. A policy of disarmament is impeccable in terms of deontology and motives. However, in consequentialist terms, its failings are so severe that relatively few groups or individuals seriously advocate this path in an age of superpower rivalry.

The moral strength of offensive deterrence built on the assurance of nuclear retaliation is summarized in the assertion, "Deterrence by fear of massive destruction has prevented, and will prevent nuclear war; since the threat of a nuclear holocaust prevents the holocaust, the threat is moral. Any posture that undermines deterrence makes nuclear war more likely, and, in consequentialist terms, is immoral." Nevertheless, though an offensive standoff may satisfy one moral criterion—it deters war—the means by which it achieves that end—a threat to kill millions of innocents—is morally unacceptable.

A policy of defense dominance can, in theory, resolve the moral dilemmas, in deontological terms, associated with threatening massive numbers of innocent citizens. A strong defensive standoff deters war by denying an aggressor any hope of achieving his military objectives. Defensive postures therefore deter war without threatening immoral acts. One would consequently have expected strategic analysts, no less than religious leaders and philosophers of ethics, to embrace the defensive posture. That this has not happened—indeed, most discussions of morality and security ignore the defensive posture entirely—can be attributed in large part to the following:

1. Some strategists fear that the defensive posture that deters by denial of aggressive intent is simply not as powerful a deterrent as is the fear of nuclear retaliation.[2]

2. Many analysts, perhaps a majority of them, believe that there simply is not and cannot be an effective defense against nuclear weapons. The image of nuclear weapons as "destroyers of civilization" or "ultimate weapons" is strong. Obviously, moral issues are moot if a defensive posture cannot be achieved technically.[3]

3. The prospect of either superpower willingly substituting defenses for offensive strength is labeled utopian and, therefore, not worthy of further investigation.[4]

4. The potential application of defenses is frequently derided as a technical fix to a political problem.[5] Certainly, the construction of defenses, in conjunction with increasing offensive weaponry and intense political rivalry, is a futile and potentially dangerous initiative; such a policy however, constitutes just one possible course of action.

TABLE 1. Strategic nuclear force structures as of 1987[7]

	United States	Soviet Union
Land-based		
Launchers	1,000	1,392
Warheads	2,310	6,846
Sea-based		
Launchers	640	928
Warheads	5,632	3,232
Air-based		
Launchers	361	155
Warheads	5,079	1,170

5. Many fear that the actual transition from offense dominance to defense dominance will be fraught with danger and instability.[6]

None of the arguments listed above, even when taken together, justify the omission of a defensive world in discussions of morality in the nuclear world. What we intend to describe is, first, a defensive posture that meets the above objections and retains the moral superiority that so many philosophers find appealing and, second, a way to move from today's offensive world to tomorrow's defensive world.

The MAD Era

We are now in the thirty-ninth year of the MAD era, the Soviet Union having exploded its first atomic bomb in September 1949. The two superpowers each possess more than 10,000 strategic nuclear warheads, most of which are equivalent to the explosive force of more than 50,000 tons of TNT.[7] The most accurate delivery systems are the land-based ICBMs. Each side also possesses submarine-based ballistic missiles, as well as strategic bombers armed with cruise missiles and gravity bombs. I summarize the strategic arsenals of the two superpowers as of 1987 in Table 1.

The very accurate ICBMs are regarded as first-strike weapons, since they are able to take out ICBM silos. In principle, the Soviets' 6,846 ICBM-launched warheads could take out all 1,000 of the U.S. ICBM silos; on the other hand, the 2,310 U.S. ICBM-based warheads are insufficient to destroy all the 1,392 land-based Soviet ICBMs, since at least two weapons must be assigned to each target. Some analysts regard this situation as compromising the principle of mutually assured destruction: The Soviets could wipe out the U.S. *accurate* ICBM force with a first strike, but the United States could not do the same if it struck first.

I find this argument unconvincing because, even after such a blow, the U.S. submarine- and air-based nuclear arsenal is sufficient to obliterate most of the Soviet Union's industrial infrastructure, as well as key military centers. Thus, even though the declared policies of both superpowers reject attacks on industrial and civilian facilities, ultimately and *in extremis*, it is exactly this kind of attack that each side threatens; countervalue, though inadvertent, is a linchpin of offensive-deterrent doctrine.

What I have said about the superpowers holds even more strongly for the other weapons states: the United Kingdom, France, the People's Republic of China (PRC), and India (which has exploded a nuclear device but denies being a weapons state). Since these states do not have enough weapons to launch a counterforce attack, their targeting doctrine is countervalue. They deter attack by explicitly threatening retaliation against cities.

Finally, there are the implicit nuclear states: Israel, probably South Africa, Pakistan, and perhaps others. These have not exploded bombs, but they too regard their possession of nuclear weapons, even untested ones, as effective deterrents that pose a threat of dire punishment to attackers (mainly to civilians?) in the event their own national existence is threatened.

As the MAD era passes the end of its fourth decade, we are not yet quite Wohlstetter's nuclear-armed crowd.[8] Yet there are, in addition to the U.S.-Soviet nuclear confrontation, several other arenas of nuclear conflict—for example, U.K.–USSR; France–USSR; PRC–USSR; PRC–India; India–Pakistan; Israel–Pakistan. And despite denials expressed in reassuring declaratory policies, can one doubt that the fear of retaliatory horror wreaked on civilians, no less than military forces, operates as the ultimate sanction against nuclear war?

The moral sanction of the MAD era has been consequentialist: MAD has prevented nuclear war so far. We cannot prove that MAD, even among Wohlstetter's nuclear-armed crowd, would be less effective in deterring war than it has been in respect to superpower confrontation. If the MAD era fares poorly as judged on deontological grounds, I have no doubt that it would still be preferred were it the only posture that deters war. Thus, when we seek alternatives that theoretically are less burdened by abstract deontology, we must be confident that the alternatives are as effective as MAD in maintaining the peace.

In judging the moral acceptability of MAD, we cannot ignore the moral asymmetry between liberal democracy and totalitarian communism. The extraordinary events now taking place in the Soviet Union—with, for example, Gorbachev proposing to erect a monument to the memory of the victims of Stalin's purges or a limitation on the tenure of the Soviet governmental leadership—suggest that the moral asymmetry may gradually approach moral symmetry. But we cannot say this with any certainty; and as long as the Soviet system maintains its historical expansionism, the liberal democracies cannot afford to trade a strategic posture that has worked for a more moral posture that might fail.

The MAS Era

An alternative to MAD is MAS, in which offensive weapons have been greatly reduced and are dominated by defensive systems. This defense-dominated world was first proposed by Herman Kahn[9] in 1967. In this world, the offensive weapons are reduced not to zero but to what we call the "cheating threshold" (CT)—that is, the number of weapons that could conceivably be produced and sequestered clandestinely, given advanced national and multinational capabilities for verification. The number of weapons in an acceptable CT would be the subject of intense superpower examination and discussion.

I cannot imagine nuclear weapons disappearing altogether, at least while the world consists of states, some of which are liberal democracies, some illiberal totalitarianisms. No state, least of all a superpower, will be willing to rid itself of all offensive weapons, even if it possesses impregnable defenses. The possibility of cheating or blunder is all too real. Thus, in some sense, but to a very much attenuated degree, fear of offensive retaliation will operate even in the MAS era. On the other hand, the retaliation that is threatened when a few hundred offensive missiles face huge defenses is by comparison much less catastrophic than when 10,000 warheads face no defense, as is the case in today's MAD world.

The size of the CT depends on the efficacy of surveillance methods. (Again, the mutual surveillance techniques being developed for implementing the INF treaty suggest that monitoring may be much more intrusive than we would have believed possible a few years ago.) The

CT will also depend on the effectiveness of the defense: As defense becomes more effective, the tolerable cheating threshold could be increased. For the sake of argument, we will consider the CT to be about 200 single-warhead ICBMs. This is of the same order as the 400 nuclear weapons President Eisenhower insisted were sufficient in 1956.

We would envisage the MAS posture to require deployment of "magnificent" defenses—a response sufficient to neutralize 200 weapons. At this low attack size, the defenses need not be exotic; they can use kinetic-kill vehicles (KKV, or "smart rocks") and can be ground based. In particular, I envisage a two-layer, ground-based system. The top layer attacks reentry vehicles (RVs) in late mid-course with exoatmospheric reentry interceptor system (ERIS) defenders; the bottom layer attacks RVs in terminal course with high endoatmospheric interceptor (HEDI) defenders. This system makes up the bottom two layers of the three-layer system analyzed by the George C. Marshall Institute.[10]

The George C. Marshall Institute has claimed that a three-layer system using only KKVs would be 93 percent effective against a threat of 11,200 warheads and 90,000 decoys. I estimate that a two-layer system would destroy on average all but 6 warheads of a threat consisting of 200 RVs and 1,600 decoys. Spergel and Field have vigorously attacked the Marshall estimate of effectiveness of its space-based defense against the 11,200 RV threat;[11] however, this criticism does not affect the estimates of the effectiveness of the bottom two layers, nor is it relevant to the case of a very small attack.

In his recent book, Simon Ramo, the developer of the original Atlas ICBM, suggests that a ground-based kinetic-kill system would be 90 percent effective against 1,000 RVs (that is, 10 percent would escape destruction).[12] If we assume that the effectiveness of the system increases linearly as the threat diminishes (0 percent "escaping" when the threat is zero), then we might argue that only 2 percent would escape destruction if the threat is reduced from 1,000 to 200. This leads to 4 warheads, on average, landing on target—a number close to the one derived from the Marshall estimate.

Finally, we mention that the workshop on SDI held at the Evangelical Academy Tutzing in 1986, concluded that "any SDI system realistically conceivable will be able to handle not more than 200 arriving reentry vehicles." This statement is based on analysis presented by former Assistant Secretary of Defense Robert S. Cooper.[13]

I now consider the moral sanctions of MAS. First, with respect to means—that is, as judged deontologically—MAS is at first sight impeccable. As President Reagan argued in his Star Wars speech, it is incomparably better to protect against (and thus deter) a nuclear attack than to avenge a nuclear attack. And in the MAS era, with *both* sides possessing superb defenses, the moral case for MAS—that its means threaten no innocents—is unassailable.

On the other hand, a one-sided deployment of defense is not morally impeccable. A defense that operates magnificently against a ragged retaliation of a few hundred RVs might not be effective against a full strike of 10,000 RVs. Thus, if only one side has defense, one encounters a situation of "conditional survival" (to use the language of Kent and DeValk).[14] A first strike is effective since the defense can handle the retaliatory ragged response; survival is thus conditional on striking first. Thus, defense per se can be morally impeccable or morally unacceptable, depending on the circumstances. Of course, we in the West cannot conceive of America striking first, and so this moral deficiency of SDI has an unreal aspect to it. On the other hand, we cannot deny that unilateral SDI can appear threatening as perceived by the Soviet Union.

But this is irrelevant to the morality of a MAS era, in which both sides will have shifted to defense dominance. In that event, the difference in strength between a first strike and a second, retaliatory strike disappears, leaving no incentive to strike first. The presumed moral deficiency if only one side possesses defense might arise in the transition from MAD to MAS, but it is not relevant to the analysis of the fully established MAS era.

In the MAS era, the question of motives ought to be less troublesome than in the MAD era—or, perhaps more accurately, if in the MAS era strategic nuclear weapons become obsolete, the question of motives can hardly bear on the moral sanctions of MAS. One hopes that neutralizing nuclear weapons will lead to a general amelioration of international relations and a mollifying of aggressive tendencies—but this is a hope that we cannot be confident will materialize.

The consequentialist sanction for MAS is its moral Achilles' heel. Would the prospect of launching a futile offense against a superb defense actually deter nuclear aggression in the absence of any credible threat of immediate retaliation? We know, empirically, that absence of nuclear war and MAD have been perfectly correlated for the past forty years; though this does not prove that MAD prevents nuclear war, neither is it inconsistent with this supposition. By contrast, we have no empirical evidence to support the effectiveness of MAS to deter nuclear war.

All we can say is that, if a defense clearly overwhelms an offensive threat, a potential aggressor would perceive little advantage in instigating nuclear war. At the very least, defenses inject an element of uncertainty in the mind of a would-be button pusher, compounding the uncertainties already inherent in a nuclear attack against an undefended, but nuclear-armed, opponent.

If, in the MAS era, nuclear weapons become obsolete, what happens to the doctrine of extended deterrence, the ability of the United States to threaten nuclear retaliation for conventional Soviet incursions in Western Europe? The simple answer: extended deterrence is not credible in the MAS era. But has extended deterrence ever been a realistic policy? Can we really expect U.S. presidents to launch a nuclear attack in defense of Berlin or Hamburg? And the INF treaty all but implies that extended deterrence, in this sense (as opposed to tactical nuclear weapons), may have been bluff anyhow.

The demise of extended deterrence may turn out to be less serious than some Europeans believe because of technical and political developments. Precision guided missiles and similar devices hold the potential for creating a nonnuclear force able to counter the advantages in manpower and weaponry now held by the Soviets. One would hope that, in the MAS era, defenses against shorter-range nuclear devices—tactical as well as strategic—might be possible. Though Europe's proximity to the East diminishes the transit time of nuclear weapons, European defenses need cover a much smaller area than, say, the continental United States. To some extent, this lesser area of vulnerability may compensate for the shorter flight times of attacking RVs. I realize that these are speculations, perhaps optimistic speculations. Yet one cannot ignore the signals now coming from the Soviet Union calling for mutual conversion of the conventional forces on both sides from offensive to defensive postures and weapons. One can only hope that these political overtures are real and that they presage a Soviet renunciation of militant, expansionist communism.

Finally, we consider the relative stability of MAS and MAD in respect to technological breakthroughs and cheating. C. L. Glaser has argued that the MAD posture, with 10,000 undefended RVs on each side, is robust against cheating or even technological improvements: 500 clandestine RVs hardly change the balance, nor does a 50 percent higher yield on each warhead.[15] By contrast, he argues, a defensive confrontation is less robust: for example, if the defense is 95 percent effective against 200 RVs, allowing leakage of 10 warheads, it might be only 90 percent effective against 300 RVs, allowing leakage of 30 warheads. The difference between 10 and 30 warheads is regarded as important and might lead to instability in perception. On the other hand, were the defense much better—say 99.9 percent effective against 200 RVs and 99.8 percent against 300 RVs—the difference is far less alarming: 0.2 RV against 0.6 RV. Thus, an extremely good defense is adequate protection against an enlarged (presumably clandestine) offense, just as a very large offense will offset an incremental increase in offense. This, however, is not to say that a breakthrough in ballistic missile defense would not be destabilizing—after all, Israel neutralized the entire Syrian Air Force, presumably by radar jamming, during its Lebanon incursion. All one can say is that the defense visualized here is multilayered—and overwhelming. Its neutralization is not easy.

We summarize our reflections as follows: MAD's main moral sanction is consequential, its main moral weakness is deontological; by contrast, MAS's main moral sanction is deontological, its possible, though by no means demonstrated, moral weakness is consequential. As for motives, one can hardly imagine a MAS era in which motives have not been altered for the better; by contrast, the present MAD confrontation exists, as much as anything, because neither side accepts what it regards as the other's motives.

Managing the Transition from MAD to MAS

Is it possible to conceive of strategies of defensive deployment that are stable—that is, that do not increase the incentive to strike first, nor lead to an escalating arms race? Doubts as to the possibility of such a transition, even in principle, have generally been invoked by opponents of defensive deployment, especially SDI. Yet, in truth, remarkably little analytical effort has gone into devising strategies for managing the transition to defense. Here, we shall briefly analyze the stability of the transition.

The main idea underlying all of the schemes for achieving a stable transition from MAD to MAS is what Jack Barkenbus and I have called the Defense-Protected Build-down (DPB): Every deployment of a defensive missile must be compensated for by the removal of an offensive missile by the side deploying the defense.[16] The basic idea is that if a side is *not* intending a first strike, but is counting instead on its ICBMs simply as retaliatory weapons, then it needs fewer defended than undefended ICBMs to maintain a given level of retaliatory threat.

For example, if side A has 1,000 undefended ICBMs and expects to lose 500 of them if side B strikes first, it still could deliver a 500-

weapon retaliatory strike. Were side A to deploy Ballistic Missile Defense (BMD) that guaranteed survival of 500 defended ICBMs even if side B strikes first, side A could dismantle 500 of its ICBMs and still maintain its ability to retaliate with the remaining 500 ICBMs. The notion that one needs fewer defended than undefended ICBMs to maintain a given retaliatory capability was first analyzed in detail at the Los Alamos National Laboratory in 1981 during the debate over deployment of the MX missile.[17]

We envisage DPB to be a reciprocal process that can be initiated either unilaterally or cooperatively. Thus, side A, equipped with an initial force of 1,000 ICBMs, might begin the process by deploying 50 interceptors and simultaneously removing 50 ICBMs; this action signals to B that side A does *not* intend to strike first. Side B can therefore, with impunity, remove somewhat more than 50 ICBMs (since it now faces only 950 instead of 1,000 offensive weapons) at the same time it deploys its defense. This reciprocal sequence of reduction in offense and deployment of defense can be repeated until the process stabilizes at a low level of offense—the limit will be determined by the effectiveness of the defense.

As visualized, at every level of the DPB process, only missiles, not cities, are defended. Were cities as well as missiles defended, the side that strikes first would gain the advantage: Its defense, which might be insufficient to handle a first strike against its cities, might be sufficient to handle a ragged, retaliatory second strike. On the other hand, were cities left undefended, MAD would operate at each level of the DPB: Fear of retaliation against cities would remain the (morally unappetizing) sanction against outbreak of nuclear war. Thus, in this highly simplified version of DPB, defense is deployed to protect the retaliatory force alone. We shall consider defense of cities, as well as defense against slower offense systems (such as bombers and cruise missiles), subsequently.

Roy Radner has given a mathematical analysis of this simple DPB process.[18] In his analysis, a side is deterred from attacking if, even after the attacker uses *all* of his weapons to eliminate the defender's ICBMs, the defender still has a "desired reserve" (designated as s) with which he can retaliate against the attacker's cities. Any configuration in which both sides meet this criterion is in the "zone of mutual deterrence." Radner shows that a sequential build-down of counterforce offense and buildup of hardpoint defense can be programmed so as to maintain the confrontation in the zone of mutual deterrence at every stage of the build-down.

TABLE 2. Defense-protected build-down process from initial position of 5,000 warheads on each side. (Defense effectiveness, $q = 0.6$, desired reserve, $s = 100$, desired level of deterrence = 0.999.)

Round	Side 1	Side 2
0	5,000	5,000
1	2,207	1,054
2	571	365
3	275	235
4	217	209
5	206	204
—	—	—
Limit	203	203

Radner's approach is probabilistic—that is, at every level of defense effectiveness q, no matter how good, there is some probability that the attacker destroys all of the defender's strategic reserve, provided the number of attacking warheads is large enough; conversely, this probability might be extremely small if the defense is very effective ($q \approx 1$) and the number of missiles is very small. Thus, Radner introduces the concept of the "level of deterrence," δ—that is, the probability in a particular offensive–defensive confrontation that an attack will not succeed in destroying the defender's strategic reserve. For example, if this probability is 0.999, the attacker is "deterred at a level of deterrence $\delta = 0.999$."

Radner gives several examples of how a defense-protected build-down process converges. In Table 2, we give Radner's results for a DPB process in which each side starts with 5,000 non-MIRVed warheads, each increases its defense effectiveness to $q = 0.6$ (that is, defense destroys on average 60 percent of incoming warheads), the desired reserve is 100 warheads, and the level of deterrence is $\delta = 0.999$.

As we see from the table, each side needs only $\sim$203 warheads to assure, with probability 0.999, that it can ride out a full-scale attack and still have 100 retaliatory weapons.

In any reciprocal bargaining over defense/offense, each side will tend to underestimate the effectiveness of its own defense and overestimate the effectiveness of its opponent's defense. Thus each side will claim it needs to dismantle fewer offensive weapons to compensate for its deployment of defense than its opponent claims it must. Can we

TABLE 3. Limiting offensive deployment as a function of defense effectiveness, q, and desired reserve, $s = 100$, initial position of 5,000 warheads on each side.

q	Limiting offensive deployment
0.1	1,341
0.2	659
0.3	432
0.4	318
0.5	249
0.6	203
0.7	169
0.8	144
0.9	123

resolve this discrepancy in perceived "compensation-price ratio?" Indeed, critics of the whole concept of DPB, or of negotiated defense deployment within an arms control agreement, consider this asymmetry in perception to be a fatal flaw of DPB.

Radner's analysis suggests a way out of this dilemma. In Table 3, I quote Radner's limiting deployments of offense, from an initial 5,000 warheads for different defense effectiveness q, a desired reserve of 100, and a level of deterrence $\delta = 0.999$.

The effect of the aforementioned asymmetry in perception is seen in Table 3. If one side claims its defense is 0.3 effective, it should be prepared to build down to 432 offensive missiles; if it claims the other side's defense effectiveness is 0.6, it will be prepared to allow only 203 offensive missiles for its opponent. A compromise between these estimates might be their average—318 missiles. A 318-missile attack against a $q = 0.3$ defense would allow about 96 retaliatory weapons; against a $q = 0.6$ defense, it would allow about 191 retaliatory warheads to survive. Although the ratio $191/96 \sim 2$ appears formidable, the situation is still one of mutually assured destruction: 96 retaliatory weapons are a serious threat, practically no less than 191!

Radner's ingenious analysis suggests that DPB in principle can be achieved without encountering unacceptable crisis instability, even in the face of asymmetric perceptions of defense effectiveness. Nevertheless, one must concede that his analysis, confined as it is entirely to missile defense, is too schematic. We therefore turn to more elaborate analyses of stable defensive deployments in which other offensive modes (bombers, cruise missiles) and other targets (cities, military installations) are taken into account.

More General Defensive Deployments

Four studies—G. C. Reinhardt's *Defense and Stability*,[19] D. Wilkening and K. Watman's *Strategic Defense and First Strike Stability*,[20] G. A. Kent and R. J. DeValk's *Strategic Defenses and the Transition to Assured Survival* (see footnote 14), and S. O. Fought's *SDI: A Policy Analysis*[21]—purport to trace theoretically stable paths from MAD to MAS.[22] All of these studies except Reinhardt's explicitly consider city and silo defense. In this respect, they are more focused on the MAD–MAS transition than Radner; Reinhardt considers city defense alone.

To anticipate and summarize their findings, all the authors (except Reinhardt) agree on the following basic elements of a crisis-stable path to MAS.

Step one: Deploy silo defense, either active or passive, and reduce counterforce ICBMs, either by arms control or DPB. During this phase of the transition, nuclear war is deterred by fear of destruction of cities. During this period, the conditions of Radner's analysis are fulfilled (defended ICBMs, undefended cities), and, as he showed, both sides are in his region of mutual deterrence.

Step two: Deploy area, or city, defense only after counterforce threats on both sides have been greatly reduced. The rationale for this strategy is the following: Crisis instability occurs if either or both sides perceive an advantage in striking first; that is, if a first strike renders the retaliatory strike "ragged" and capable of being thwarted by defenses that are insufficient to handle a full first strike. But if the silos are invulnerable, then the retaliatory strike is no longer weaker than the first strike. An attacker no longer gains advantage by striking first, and so he no longer has incentive for such action.

Step three: Deploy air defense only after steps one and two have been accomplished. (Step three is dealt with only in the Wilkening paper.) Since air-breathing missiles are slow and allow ample time for mobilizing retaliation, they are not regarded as first-strike weapons and therefore do not affect stability during a crisis.

The above-mentioned conclusions are supported by game-theoretic computations. Though the details of the analysts' computations differ, the general idea is the same. For example, in the Wilkening analysis, a numerical value is attached to each side's incentive to strike first. This first-strike incentive is calculated by comparing the following quantities: the expected damage to military targets from a first strike on the adversary and from his retaliatory strike; the expected damage from a first strike by the adversary and a retaliatory strike against his military targets; and the value to each side of having no nuclear strike (that is, preservation of the status quo for each side). If each side's first-strike incentive exceeds a given value (0.3 in the units used by the authors), the situation is regarded as unstable. Using these quantitative measures, one can map the space of U.S. defense deployment versus Soviet defense deployment and identify the areas representing stable and unstable deployments.

An example of such a map, taken from Kent and DeValk, is shown in Figure 1. Similar maps are given in all the other studies. Referring to the figure, one sees that when neither side has silo defense, the situation is crisis-stable because MAD is operative. Similarly, when both sides have perfect silo defense (upper right-hand corner), the situation is again stable because MAS is operative (although survival here means only survival of counterforce). On the other hand, if, say, the United States has deployed massive silo defense but the Soviet Union has not (lower right-hand portion of the x axis), the United States can defeat a retaliatory second strike: this situation is unstable because the United States can assure its survival by striking first (conditional survival).

In this particular example, the area in which neither side has incentive to strike first is quite wide, and a path to MAS that avoids any region of conditional survival is easily delineated. However, if assumptions other than those made in constructing the figure are allowed—for example, if area (nondiscriminating) defenses are deployed, as in Figure 2—the path from MAD to MAS is blocked by a region in which one side or the other is in a state of conditional survival. Note that the two cases differ in the kind of defense deployed.

In case one, the defense is discriminating; that is, it protects silos. In case two, it is nondiscriminating; that is, it protects cities. These two graphs thus illustrate step one in the transition from MAD to MAS: non-deployment of area defense until the ICBMs are fully protected. On the other hand, once the counterforces are invulnerable, symmetric area defense is not destabilizing. To quote Wilkening and Watman:

> Thus by completing the BMD transition first and then embarking on the air defense transition, it might be possible to arrive at a world with "perfect" defense in a stable manner (provided, of course, these hypo-

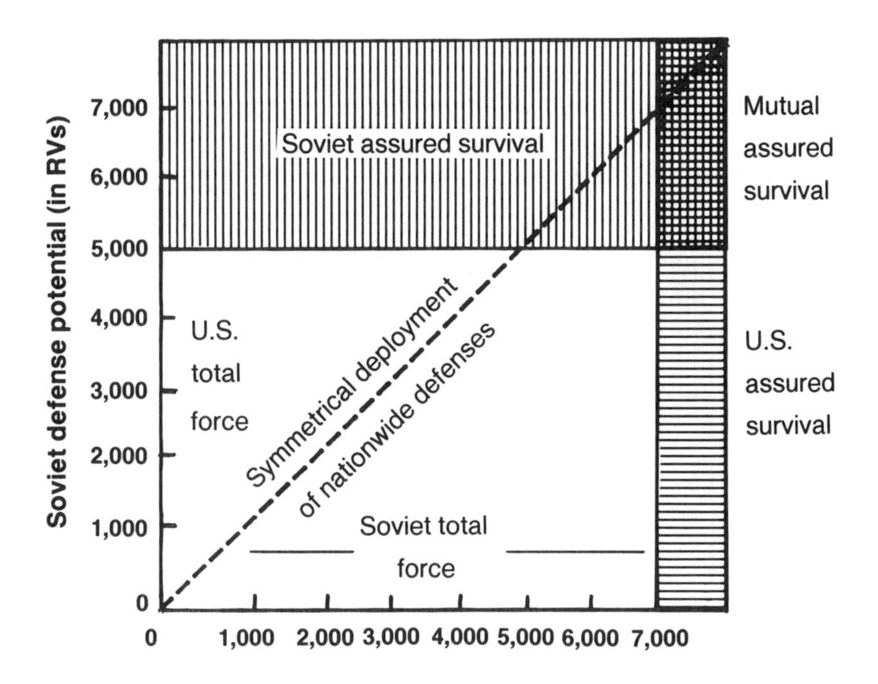

Data (National)

- Nationwide defenses operate in pure random subtractive mode and are invulnerable to suppression
- U.S. force: 5,000 on-station RVs, including
  —2,000 ICBM RVs in 1,000 silos, of which 1,500 RVs are killers with 0.4 P_k against Soviet silos
  —3,000 nontargetable RVs
- Soviet force: 7,000 on-station RVs, including
  —6,000 ICBM RVs in 1,400 silos, of which 5,000 RVs are killers with 0.7 P_k against U.S. silos
  —1,000 nontargetable RVs
  RV = re-entry vehicle
  P_k = Probability of killing a silo

FIGURE 1. *Assured survival in defense-potential format, given current offensive forces.*

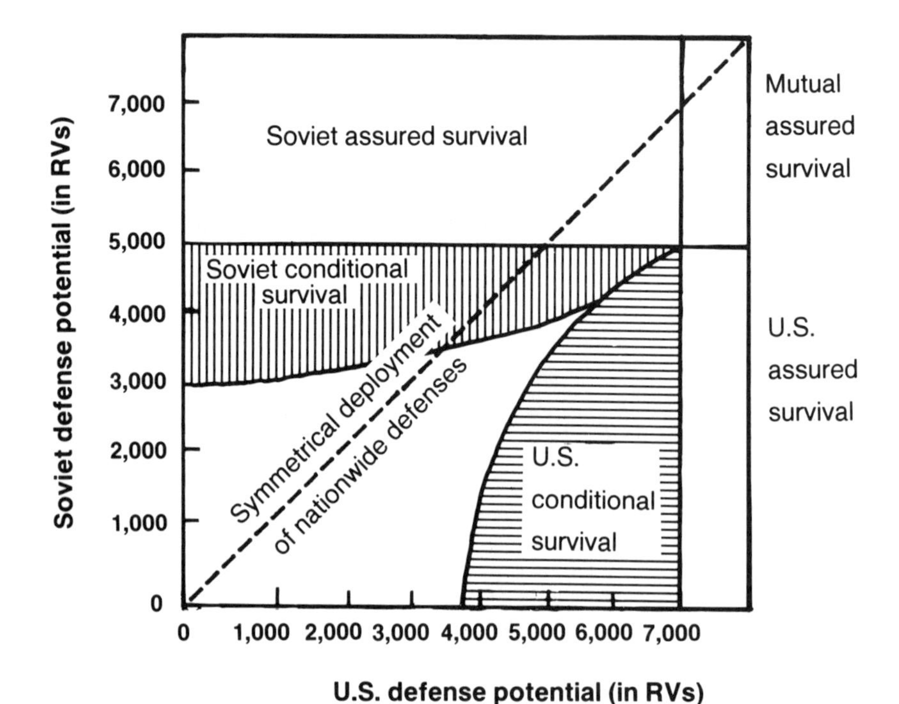

Data (National)

- Nationwide defenses operate in nondiscriminating random subtractive mode and are invulnerable to suppression
- U.S. force: 5,000 on-station RVs, including
  —2,000 ICBM RVs in 1,000 silos, of which 1,500 RVs are killers with 0.4 P_k against Soviet silos
  —3,000 nontargetable RVs
- Soviet Force: 7,000 on-station RVs, including
  —6,000 ICBM RVs in 1,400 silos, of which 5,000 RVs are killers with 0.7 P_k against U.S. silos
  —1,000 nontargetable RVs
  RV = re-entry vehicle
  P_k = Probability of killing a silo.

FIGURE 2. *Base Case II: conditional survival, given current U.S. and Soviet ballistic missile forces and nondiscriminating strategic defenses.*

> thetical "perfect" defenses are technically achievable and survivable). Therefore, to minimize instabilities during the defense transition, terminal BMD, nationwide BMD, and nationwide air defense should be deployed sequentially. . . . Cooperation of both sides will be required to effect this phased transition.

Limitations of Game-Theoretic Analysis

How seriously should we take quantitative analyses of the stability of defensive transitions? Obviously, these analyses are almost grotesquely schematic. No one, especially those who perform such analyses, believes that policymakers decide whether or not to launch a nuclear war on the basis of whether a "game value" is larger or smaller than 0.3. To quote from the critique by C. Max *et al.*, "Quantitative studies of the genre discussed below can be genuinely misleading if taken literally. The accompanying assumptions are so restrictive as to severely limit the applicability of such studies."[23]

Despite such skepticism, these studies are useful in delineating areas of concern if and when the two sides proceed toward a defensive transition. The essential sequencing of crisis-stable steps toward MAS is robust. These steps are: (1) deployment of silo-defenders coupled with compensating reductions in first-strike RVs (that is, some version of DPB) until the first strike forces on both sides are invulnerable; (2) gradual and reciprocal deployment of city defense against ICBMs; and (3) defense against slow (air-breathing) devices. This sequence depends on common sense, not upon game-theoretic analysis. The fact that analysis supports common sense simply demonstrates that game theory, if *not* used literally, is not misleading.

The main conclusion of these analyses, as well as common sense, can be termed an "existence theorem": there exist plausible, crisis-stable paths leading from MAD to MAS. These paths obviously require a high degree of cooperation between the two parties, but as the INF treaty suggests, the type of cooperation that seemed to be entirely out of the question a few years ago now (1992) is within reach. In the remainder of this paper, I shall therefore assume that a MAS world has been achieved, and I will consider the ultimate role of nuclear weapons in such a world.

The Era of Nuclear Renunciation

We live year by year, yet the nuclear threat forces us to contemplate a world centuries, even millennia, into the future. The knowledge of how to destroy the world can never be erased from human consciousness. Just as each of us must live out his life with the knowledge of how to end that life (by jumping from a ten-story building, for example), so, at least since the detonation of the first thermonuclear bomb, we must live with the knowledge of how to eliminate our entire species. Most people do not commit suicide, although some do. Can we do as well collectively as most of us do individually? Can we exorcise suicide by thermonuclear Armageddon, in spite of possessing ineradicable knowledge of how to accomplish that feat?

Most of us do not commit suicide because we abhor death and almost always seek ways to avoid it. This choice is a personal one over which each of us has control. In principle, however, there is another technical approach to preventing suicide: to banish all of the instruments that might be employed in committing suicide. This, of course, is preposterous—even if all handguns were destroyed, one could not prevent a would-be suicide from drowning himself. So one must conclude that only because of a powerful psychological inhibition—our instinctive distaste for death—do we as individuals avoid suicide.

This analogy to mankind's facing suicide through thermonuclear holocaust is not entirely far-fetched. We have avoided nuclear war thus far because we abhor death—in short, because each side knows that resort to nuclear attack would be suicidal. In other words, MAD has worked, yet we may not be able to count on MAD working forever. Thus, we have proposed MAS as a morally superior posture, which ought also to deter nuclear war.

But would MAS itself be good enough over the millennia we here contemplate? After all, the MAS era, with its offensive arsenals kept at the cheating threshold, hardly eliminates nuclear weapons. We still must contemplate a world subject to nuclear terrorist attack—megaton bombs secretly concealed in the World Trade Center towers in New York City or the Rossiya Hotel outside the Kremlin. To contemplate the shape of a world that is truly free of nuclear threat is not an idle exercise, even if today we face the possibility of massive nuclear destruction.

Can we visualize the conditions necessary to achieve a world free of nuclear threat; or at least, can we visualize a world in which nuclear threat is far less serious than we now deem it to be? I would propose two necessary preconditions for such a world: (1) Kant's comity of liberal democracies, and (2) The sanctification of Hiroshima.

Kant's comity. Achievement of a fundamentally peaceful world is a utopian vision that has engaged philosophers over the ages. Immanuel Kant's idea for a peaceful comity of nations, propounded 200 years ago, is still the best vision: If the world consisted of liberal democracies, in which the people who must do the fighting and dying decided whether or not to go to war, there would be no more war.

Kant's thesis has been subjected to empirical test by political scientists in the past few years. For example, R. J. Rummel has concluded that, in the past hundred years, no war has been fought between two genuine liberal democracies.[24] I shall assume that Rummel's finding gives strong support to Kant's idea that a comity of liberal democracies is peaceful.

Of the world's 140-odd nation-states, only a small minority, perhaps 20, are truly liberal democracies. Some might think it fatuous to speak of most of the world becoming liberal-democratic. Nevertheless, recent developments in the Soviet Union and Eastern Europe suggest that, perhaps in the next century if not in this one, democratic, open values will be universally recognized as superior to any alternative.

The Sanctification of Hiroshima. Let us suppose Kant's comity has been achieved. Would defensive systems still be needed? After all, the OECD states, though economic rivals, are surely not aiming offensive nuclear weapons at each other. If, in the next century, the relations between all the major nations become friendly, why would we need any nuclear weapons? In the utopia I outline, I don't think we will need nuclear weapons—we may achieve Jonathan Schell's abolition[25]—yet the knowledge of the weapons will exist. And that knowledge alone might serve as an implicit deterrence: Should a non-nuclear war start despite Kant's strictures, and if national survival were at stake, the war might end with exchange of thermonuclear bombs.

This "implicit deterrence" may be sufficient to prevent conventional war (as may have been the ultimate driving force in the Israeli–Egyptian rapprochement). From this point of view, the possibility of the ultimate use of nuclear bombs may be a desirable sanction.

But there is another way to look at the matter. Suppose nuclear weapons disappeared from the Earth. Would the world be better or worse off? I would say better, even though the "implicit deterrence" implied in the knowledge of how to rebuild nuclear arms would remain. In making this assertion, I assume not only that the world is much more peaceful than it is today but also that conventional defense technology can deter conventional attack. From this point of view, we ought to seek ways of banishing nuclear weapons, in the same sense that, according to some traditions of desert warfare, the poisoning of water wells is taboo. Can nuclear weapons be surrounded by such taboos?

One approach to establishing powerful taboos against nuclear weapons may be suggested by the outpouring of concern at the fortieth anniversary of Hiroshima in 1985. This outpouring of emotion by hundreds of thousands of participants far surpassed that at previous observations of the Hiroshima bombing. Indeed, it seems that, as the years go by, Hiroshima, instead of gradually receding from our collective consciousness, is becoming more strongly impressed upon it.

Are we witnessing a gradual sanctification of Hiroshima—that is, the elevation of the bombing of Hiroshima to the status of a profoundly mystical event, an occurrence ultimately of the same religious force as biblical events? I suspect that the fortieth anniversary of Hiroshima, with its vast outpouring of concern, huge demonstrations, and wide media coverage, bore a resemblance to the observance of major religious holidays and that, indeed, Hiroshima is becoming sanctified.

The sanctification of Hiroshima may be a hopeful development for the nuclear era. How can mankind accept, at the most fundamental level, the absolute necessity of avoiding nuclear holocaust—50, 100, 1,000 years after Hiroshima—unless Hiroshima becomes a ghastly legend, accepted by all and as well known to all as the Crucifixion is known to Christians, as Cain's slaying of Abel is known to the Jews, and as the Hegira is known to Muslims? In short, only by sanctifying Hiroshima can we expect its lesson to be learned and passed on forever.

Historical events generally fade from human memory; few of us remember, for example, what happened at the battles fought by Genghis Khan. Yet, if the awful knowledge of the effects of a thermonuclear explosion is to persist and to deter destruction for millennia to come—during which, it is hoped, no nuclear explosion will take place—then the knowledge of Hiroshima cannot be allowed to fade.

How can memory of an event be fortified and refortified over such long periods? According to Peter L. Berger, only by sanctifying the event; that is, by converting the historical event into a religious celebration.[26] To quote from Berger:

> Men forget. They must, therefore, be reminded over and over again. Indeed, it may be argued that one of the oldest and most important prerequisites for the establishment of culture is the institution of such "reminders," the terribleness of which for many centuries is perfectly logical in view of the "forgetfulness" that they were designed to combat. Religious ritual has been a crucial instrument of this process of "reminding." Again and again it "makes present" to those who participate in it the fundamental reality-definitions and their appropriate legitimations. The farther back one goes historically, the more does one find religious ideation (typically in mythological form) embedded in ritual activity—to use more modern terms, theology embedded in worship. . . . The performances of the ritual are closely linked to the reiteration of the sacred formulas that "make present" once more the names and deeds of the gods. . . . Both religious acts and religious legitimations, ritual and mythology, *dromena* and *legoumena*, *together* serve to "recall" the traditional meanings embodied in the culture and its major institutions. They restore ever again the continuity between the present moment and the societal tradition, placing the experiences of the individual and the various groups of the society in the context of a history (fictitious or not) that transcends them all. It has been rightly said that society, in its essence, is a memory. It may be added that, through most of human history, this memory has been a religious one.

Not every historical event can become a religious event. As the late Rev. William Pollard pointed out,

> The theologian and scholar Mircea Eliade has made a distinction between a people's "profane time" and its "sacred time." In sacred time, historical deeds and events gradually come to partake of the permanence of myth, while in profane time they gradually lose their grip on people and become merely historical material for historians; i.e., they remain in historical time. Is this the destiny of Hiroshima—to become a universal myth deeply grounded in the sacred time of all the peoples on earth, the symbol of their conviction that nuclear warfare must never be allowed to occur?[27]

Another characteristic of historical events that become religious events is their ability to strike awe in human minds—because of their sheer horror (as the Crucifixion) or because of their supernatural character (as the Resurrection). This first point has been made by Nietzsche: "A thing is branded on the memory to make it stick there; only what goes on hurting will stick. . . . Whenever man has thought it necessary to create a memory for himself, his effort has been attended with torture, blood, and sacrifice."[28]

Hiroshima, of course, qualifies in this respect—an event that was attended by "torture, blood and sacrifice." Indeed, this may throw a different cast on the debate about whether bombing Hiroshima was justified. Historical revisionists often claim that the bomb was unnecessary—that Japan would have surrendered readily even without the bomb. As a participant in the development of the bomb, I have never been impressed by such arguments—the simple-minded view that Hiroshima ended the war, and thereby saved perhaps a million lives that would have been lost in an invasion of Japan, has always seemed compelling to me. But from the point of view I have adopted here, I believe a deeper moral sanction justifies Hiroshima: Only an event as horrifying as that holocaust, with its burning and radiation, could be susceptible to sanctification. I cannot imagine vesting religious significance on a test; August 6, Hiroshima, not July 16, the day of the Trinity test, is remembered. This view was implied, in a sense, by Leo Szilard, the father of nuclear energy. Even before Hiroshima, he wrote: "It will hardly be possible to get political action unless . . . atomic bombs have actually been used in war and the fact of their destructive power has deeply penetrated the mind of the public."

Conclusion

The matters that I have touched upon here admit no provable conclusions. For every argument relating to nuclear war and its moral sanctions, one can adduce a plausible counter-argument. All of which is to say that there are specialists in nuclear war but no experts.

I have argued that a defensive posture is morally superior to an offensive posture. In espousing this view, I have had to assume, first, that a denial of aggressive intent is as effective in deterring war as fear of nuclear retaliation; and, second, that some variant of DPB can lead the superpowers to a mutually defensive posture without precipitating critical instability. For the first assumption, I can give no proof: The only experience so far has shown that nuclear war did not erupt when each side was able to destroy the other. For the second assumption, I adduced support from game-theoretic analysis. Such analysis is by no means infallible. I present it as much as anything to bring to the community of strategic analysts, arms controllers, and ethical philosophers, a point of view that has been unjustifiably ignored. I would hope that these considerations might put the SDI debate on a more rational basis by establishing that SDI, which almost surely cannot work against a full-fledged attack, probably makes sense against a much smaller attack. The debate ought to dwell on the moral, political, and technical advantages of defenses in a world of reduced offense, rather than on the technical feasibility of a defense against a threat so large as to be impossible to defeat.

Finally, my speculations on the sanctification of Hiroshima have little to do with today's real world. I consider them appropriate in a broad discussion of the morality of nuclear deterrence, since they bear on the question of Hiroshima's justification. The argument I give for Hiroshima's justification is neither deontological nor consequentialist —it is perhaps existentialist: Hiroshima seems to be becoming sanctified. This is a sort of existential observation. Instinctively, I believe this is a proper trend, one that should be encouraged. Oh, that I could be resurrected in a thousand years to really find out whether my instinct is correct.

References

[1] J. S. Nye, Jr., *Nuclear Ethics* (The Free Press, New York, 1986).

[2] J. Joffe, Ethics **95**, 613 (1985).

[3] See, for example, R. W. Malcomson, *Nuclear Fallacies* (McGill-Queens University Press, Montreal, 1985), p. 10.

[4] D. P. Lackey, Philosophical Forum, 1–7 (Fall 1986).

[5] *The Challenge of Peace: God's Promise and Our Response*, pastoral letter of the U.S. Catholic bishops on war and peace (National Catholic News Service, 1983).

[6] A. Carnesale, in *Ballistic Missile Defense*, edited by A. B. Carter and D. N. Schwartz (Brookings Institution, Washington, DC, 1984), p. 380.

[7] T. B. Cochran *et al.*, *U.S. Nuclear Forces and Capabilities* (Ballinger, Cambridge, MA, 1984), Vol. 1.

[8] A. Wohlstetter, *Moving Toward Life in a Nuclear-Armed Crowd* (Pan Heuristics, Los Angeles, 1976).

[9] H. Kahn, in *Why ABM, Policy Issues in the Missile Defense Controversy*, edited by J. J. Holst and W. Schneider (Pergamon Press, New York, 1967).

[10] J. Gardner *et al.*, *Missile Defense in the 1990s* (George C. Marshall Institute, Washington, DC, 1987).

[11] D. N. Spergel and G. B. Field, Nature **333**, 813–15 (1988).

[12] S. Ramo, *The Business of Science* (Hill and Wang, New York, 1988), 230ff.

[13] R. S. Cooper, in *SDI and Stability: The Role of Assumptions and Perceptions*, edited by K. Gottstein (Namos Verlagsgesellschaft, Baden-Baden, 1988).

[14] G. A. Kent and R. J. DeValk, *Strategic Defense and the Transition to Assured Survival*, RAND Report P-3369-AF (1986).

[15] C. L. Glaser, International Security, 92–123 (Fall 1984).

[16] *Strategic Defenses and Arms Control*, edited by A. M. Weinberg and J. Barkenbus (Paragon House, New York, 1987).

[17] G. E. Barasch *et al.*, *Ballistic Missile Defense: A. Potential Arms-Control Initiative*, Los Alamos Report LA-8632 (January 1987).

[18] R. Radner, in Weinberg and Barkenbus, *op. cit.*, 111–43.

[19] G. C. Reinhardt, UCRL-53743 (Lawrence Livermore National Laboratory, 1986).

[20] D. Wilkening and K. Watman, Survival, 137–65 (March/April 1987).

[21] S. O. Fought, *SDI: A Policy Analysis* (Naval War College, no date).

[22] All four of these studies, together with Radner' s analysis, are collected in Barkenbus and Weinberg, *op. cit.*

[23] C. Max *et al.*, JASON/MITRE Report no. JSR-85-926 (McLean, VA, 1985).

[24] R. J. Rummel, Journal of Conflict Resolution **27**, 27–71 (1983); M. W. Doyle, American Political Science Review **80** (4) (1986).

[25] J. Schell, *The Abolition* (Alfred A. Knopf, New York, 1984).

[26] P. L. Berger, *The Sacred Canopy: Elements of a Sociological Theory of Religion* (Doubleday & Co., Garden City, NY, 1967), pp. 40–41.

[27] A. M. Weinberg, Bulletin of Atomic Scientists **41** (11) (1985), pp. 34.

[28] F. Nietzsche, *The Genealogy of Morals* (Doubleday, Garden City, NY, 1956), 192ff.

Part IV: Time, Energy, and Resources

The Institute for Energy Analysis was founded in 1974; I served as its Director from 1975 to 1985. As energy think tanks go, IEA was conservative in outlook; rather than devising energy systems that would require large changes in social structure, we sought to improve our energy systems so as to make them acceptable to the existing society. In view of my antecedents, IEA focused strongly on nuclear energy, although we did not ignore other energy systems. Our studies gradually broadened into examination of such related matters as mineral resources and the role of time in contemporary society; our findings are summarized in the essays in Part IV.

The first essay, "The Age of Substitutability," was a response to the then-fashionable pessimism of the Club of Rome. The article was prompted by the claim, in the very influential book *Limits to Growth*, that the world would soon run out of aluminum (which is the most abundant metallic element on Earth). H. E. Goeller and I, therefore, set about to demonstrate the absurdity of such claims of imminent shortages of minerals. "The Age of Substitutability" summarized our findings.

"Energy Wars" was a further reaction to the coercive utopians—the activists who see our social order, based on high technology, as being unacceptably flawed. In this article, I drew upon IEA's 1975 study "Economic and Environmental Consequences of a Nuclear Moratorium, 1987–2010." Were nuclear energy to be renounced (as it has been in the United States), we would eventually have to live in a solar world; part of our study, therefore, tried to visualize such a world and to contrast it with a nuclear world. A major semiphilosophic finding, due largely to Daniel Spreng, was that time would be valued less highly in a solar world than in a nuclear world.

"A Family of Maxwell Demons" draws upon Spreng's observation that information, as well as time, can substitute for energy. The paper also reflects the views of my old colleague and mentor, Eugene P. Wigner, on the nature of entropy. Wigner, who was closely associated with Von Neumann and Szilard, takes as a matter hardly to be argued that entropy is an anthropomorphic concept. This point, I believe, is well accepted by most physicists but continues to be disputed by engineers and, recently, by logicians such as K. Popper.

"Energy Policy and Mathematics" continues my attempt to place energy policy in a broader context—one not captured by the fashion for energy modeling of the 1970s. I had spent 1974 in Washington helping to put in place Project Independence—a $20 million effort to predict the future of the U.S. energy system. My disillusionment with this venture is evident in this article.

"Energy in Retrospect," written in 1988, brings these views up to date. I realize, in this paper, that conservation is central for *any* energy policy, but, as an old-time nuke, I remain persuaded that supply, as well as demand, remains important.

That greenhouse warming might be the main motivation for a major shift to nuclear energy became apparent very early to us at the Institute for Energy Analysis. During the 1970s we were the only group that was studying both the greenhouse effect and the social problems posed by nuclear energy. The epilogue to "Energy in Retrospect" summarizes our rationale for wide deployment of nuclear reactors as an antidote for greenhouse warming.

The Age of Substitutability

WITH H. E. GOELLER

Two conflicting views dominate our perception of our long-term future. The "catastrophists," as exemplified by the Club of Rome and *The Limits to Growth*,[1,2] believe that the Earth's resources will soon be exhausted and that society will collapse. The "cornucopians," who embrace technological optimists like me, argue that most of the essential raw materials are in infinite supply: that, as society exhausts one raw material, it will turn to lower-grade, inexhaustible substitutes. Eventually, society will subsist on renewable resources and on such elements as iron and aluminum, which are virtually inexhaustible. According to this view, society will settle into a steady state of substitution and recycling. This asymptotic state is the Age of Substitutability.

The catastrophists' view has been expounded in recent years by such economists as Kenneth Boulding[3] and Nicholas Georgescu-Roegen.[4] Their point is that, as we exhaust high-grade resources, we shall require more and more energy to extract useful minerals from more dilute sources. This trend is inexorable and ultimately must lead to a

This paper is largely based on H. E. Goeller and A. M. Weinberg, Science **191** (4228) (1976). I am grateful to Goeller for permission to use much of his work. This version appeared in *Economic and Demographic Change: Issues for the 1980's* (International Union for the Scientific Study of Population, Liege, Belgium, 1979), Vol. 1, p. 1. Reprinted by permission.

limit—where the energy required to carry on our industry simply overwhelms us.

In principle, Boulding and Georgescu-Roegen are correct. What we shall argue is that the actual limit is probably much larger than they conceive, and that, if man can develop an inexhaustible energy source, he will always have sufficient raw materials *at reasonable concentrations* to maintain an industrial civilization not so very different from the one we now enjoy.

Cornucopian economists such as H. Barnett[5] have long insisted that the marketplace would force substitutions: as high-grade ores run out, we would turn to lower-grade, more costly substitutes, and this process was seen as never terminating. But these early writers made little attempt to estimate quantitatively how much more energy would be needed, and how much more pollution would be caused, by such substitution.

It was H. E. Goeller who first pointed out that most of the mineral products our society uses—such as iron, aluminum, glass, and cement—are almost incalculably abundant; for all practical purposes, they will never "run out." The few elements that are relatively scarce—such as copper, or lead, or mercury—we can largely do without if we must; moreover, they are used on a far smaller scale than are the abundant elements, and so, even if their extraction from low grade ores entails much more energy or cost in aggregate, the total cost in energy or money would not be impossibly large.

As geologist D. F. Frasché said in 1962, "Total exhaustion of any mineral will never occur: Minerals and rocks that are unexploited will always remain in the earth's crust. The basic problem is how to avoid reaching a point where the cost of exploiting those mineral deposits which remain will be so costly, because of depth, size, or grade, that we cannot produce what we need without completely disrupting our social and economic structures."[6]

"Demandite" and "Avalloy"

A convenient way of displaying the relation between long-term supply and long-term demand of minerals is to construct a "chemical" formula for the average nonrenewable resource and the average metal that society uses. The resource we shall call "demandite," and the metal, "avalloy." The properties of demandite and avalloy (for the

TABLE 1. The average nonrenewable resource used by human society in 1968[7]: "demandite."

United States			
$(CH_{2.14})_{0.8022}$	$(SiO_2)_{0.1115}$	$(CaCO_3)_{0.0453}$	$Fe_{0.0110}$
$Al_{0.0011}$	$(Cu,Zn,Pb)_{0.0004}$		$Mg_{0.0004}$
$N_{0.0076}$	$O_{0.0053}$	$Na_{0.0053}$	$Cl_{0.0053}$
$S_{0.0023}$	$P_{0.0008}$	$K_{0.0007}$	$X_{0.0008}$
World			
$(CH_{1.71})_{0.6660}$	$(SiO_2)_{0.2117}$	$(CaCO_3)_{0.0815}$	$Fe_{0.0145}$
$Al_{0.0007}$	$(Cu,Zn,Pb)_{0.0004}$		$Mg_{0.0004}$
$N_{0.0068}$	$O_{0.0045}$	$Na_{0.0045}$	$Cl_{0.0045}$
$S_{0.0023}$	$P_{0.0007}$	$K_{0.0007}$	$X_{0.0008}$

X represents all other chemical elements.
Highest in order of demand are Mn, Ba, Cr, F, Ti, Ni, Ar, Sn, B, Br, Zr.
All others < 100,000 tons/year (world) or 30,000 tons/year (U.S.).
Values apply only to new metal from ore.
[7]Source: Derived, with some modifications, from data in U.S. Bureau of Mines Bulletin 650, *Mineral Facts and Problems* (1970).

year 1968) are summarized in Tables 1 through 4. The subscript following each chemical symbol gives the average mole fraction of that constituent.

TABLE 2. Some characteristics of "demandite."[7]

	United States	World
Total quantity, 10^6 tons	3,360.00	17,300.00
Total value, $\$ \times 10^6$	42,200.00	158,500.00
Average unit value, $/ton	12.55	9.16
Average energy used for recovery, kWh/ton	1,140.00	800.00
Total quantity, 10^6 ton moles	140.40	551.30
Average molecular weight	23.90	31.40
Per capita consumption, tons	17.00	5.00
Per capita energy, kWh	18,800.00	3,800.00
Per capita energy rate, kW	2.14	0.43

[7]Source: Derived, with some modifications, from data in U.S. Bureau of Mines Bulletin 650, *Mineral Facts and Problems* (1970).

TABLE 3. The average virgin metal used by human society in 1968[7]: "avalloy."

			United States			
$Fe_{0.8570}$	$Mn_{0.0119}$	$Si_{0.0105}$	$Cr_{0.0050}$	$Ni_{0.0015}$	$X_{0.0003}$	$Al_{0.0822}$
$Cu_{0.0138}$	$Zn_{0.0123}$	$Pb_{0.0025}$	$Sn_{0.0003}$	$Mg_{0.0021}$	$Ti_{0.0002}$	$Y_{0.0004}$
			World			
$Fe_{0.8983}$	$Mn_{0.0176}$	$Si_{0.0071}$	$Cr_{0.0045}$	$Ni_{0.0009}$	$X_{0.0002}$	$Al_{0.0447}$
$Cu_{0.0135}$	$Zn_{0.0097}$	$Pb_{0.0020}$	$Sn_{0.0003}$	$Mg_{0.0009}$	$Ti_{0.0001}$	$Y_{0.0002}$

X includes all other ferrous metals; *Y* all other nonferrous metals. Values apply only to new metal from ore.

[7]Source: Derived from data in U.S. Bureau of Mines Bulletin 650, *Minerals, Facts and Problems* (1970).

When looked at from this perspective, two remarkable facts emerge:

1. By far the largest component of demandite is reduced carbon and hydrogen (denoted by CH_x)—that is, the fossil fuels. In the United States, some 80 percent of *all* the molecules that we derive from the earth are molecules of fuel with average composition $CH_{2.14}$; for the world, which uses energy less intensively, and which uses relatively more coal and less natural gas than the United States, the fraction is 66 percent and the composition is $CH_{1.71}$. Of the

TABLE 4. Some characteristics of "avalloy."[7]

	United States	World
Total quantity, 10^6 tons	85.6	424.30
Total value, $\$ \times 10^6$	16,050.00	75,775.00
Average unit value, $/ton	187.50	178.60
Average value, $/ton, of Fe, Si, Al, Ti, Mg in avalloy	154.95	145.40
Average energy used for recovery, kWh/ton	12,300.00	11,100.00
Average molecular weight	53.7	54.8
Per capita consumption, tons	0.43	0.12
Per capita energy, kWh	5,290.00	1,340.00
Per capita energy rate, kW	0.60	0.15

[7]Source: Derived from data in U.S. Bureau of Mines Bulletin 650, *Minerals, Facts and Problems* (1970).

TABLE 5. Average mole percent composition of the earth's crust.

Crust (topmost kilometer on continents)				
$(CH_x)(\text{extractable})_{0.00004}(CH_x)(\text{unextractable})_{0.0083}$*$C(\text{oxidized})_{0.0153}$				
$O_{0.5907}$	$Si_{0.1943}$	$H\ (\text{Other})_{0.0658}$	$Al_{0.0507}$	$Ca_{0.0175}$
$Na_{0.0142}$	$Fe_{0.0132}$	$K_{0.0123}$	$Mg_{0.0120}$	$Ti_{0.0016}$
$Cl_{0.0014}$	$S_{0.0009}$	$F_{0.0007}$	$P_{0.0004}$	$Mn_{0.0003}$
$X_{0.0004}$				
where X is all other elements, and for which $X_{0.0004} =$				
$Ba_{0.000072}$	$Sr_{0.000064}$	$V_{0.000040}$	$Zr_{0.000034}$	$(Cu,Zn,Pb)_{0.000032}$
$N_{0.000027}$	$Rb_{0.000027}$	$Cr_{0.000026}$	$\text{Rare earths}_{0.000024}$	$(Co,Ni)_{0.000022}$
$X'_{0.000021}$	(X' = remaining elements)			

*"(CH_x) unextractable" represents the very large amount of hydrocarbon in the topmost kilometer of shale. Almost all of this is too dilute to extract with a positive energy balance and therefore cannot be used as a source of energy. In principle, it might be used as a source of CH_x for petrochemicals.

roughly $12/ton price of demandite in 1968, some 45 percent was contributed by the cost of fossil fuel; in 1974, this rose to 55 percent.

2. To an even greater extent, a single element, iron, dominates the composition of avalloy: 86 percent of all metal atoms used in the United States in 1968 were iron, almost 90 percent in the world. Moreover, the five components of avalloy that occur very abundantly (iron, silicon, aluminum, titanium, and magnesium) represent about 82 percent of its cost; almost all of this (95 percent) was the cost of iron.

Let us now compare the average composition of demandite and avalloy with the average composition of the earth's crust (Table 5) and the average composition of the sea and the air (Table 6). Immediately we are struck with a remarkable discrepancy between the composition of the earth and the composition of demandite, and an equally remarkable similarity between earth and avalloy. Whereas CH_x (that is, fossil fuel) represents by far the largest fraction of demandite, it is actually a rare constituent of the earth's crust (0.004 mole percent)—about as rare as neodymium! By contrast, aluminum and iron—at 5 and 1 percent, respectively—are among the most common elements in the earth's crust as well as being the dominant constituents of avalloy.

This comparison is further dramatized in Table 7, in which are given estimates of the ratio of resource to demand of nineteen of the most abundant elements. Of the thirteen most widely used elements, all but CH_x and phosphorus are essentially inexhaustible. Indeed, by far the most important scarce resource is extractable CH_x, or energy. (We can equate energy with CH_x since the two are interchangeable: CH_x is a source of energy, and energy, in combination with carbon dioxide and water, is a source of CH_x.) Shortages of almost all other minerals are of second order compared with shortages of CH_x.

The Principle of Infinite Substitutability

We now state the principle of "infinite" substitutability: With three notable exceptions—phosphorus, a few trace elements essential to agriculture, and energy-producing fossil fuels (CH_x)—society can subsist on inexhaustible, or nearly inexhaustible, minerals with relatively little loss of living standard. Society would then be based largely on glass, plastic, wood, cement, iron, aluminum, and magnesium: whether it would be anything like our present society depends on how much of the ultimate raw material—energy—we can produce and how much energy will cost, both economically and environmentally.

To give substance to so broad a claim, it is necessary to examine in detail, not simply in principle, exactly how people could live without a particular finite resource. We do not do this here; instead, we quote from previous work on the subject by Goeller.[8]

Certainly, there are no substitutes for the elements required to sustain life. The most important of these are hydrogen, oxygen, carbon, and nitrogen; next come calcium, phosphorus, chlorine, potassium, sulfur, sodium, and magnesium, which constitute less than 1 percent of the total in living things. Besides these there are at least thirteen required trace elements: fluorine, silicon, vanadium, chromium, manganese, iron, cobalt, copper, zinc, selenium, molybdenum, tin, and iodine. Modern agriculture requires large amounts of calcium, nitrogen, potassium, and phosphorus, plus relatively small quantities of some of the trace elements. Of the major life-sustaining elements, only phosphorus is not on our list of inexhaustible elements. The average natural abundance of phosphorus in rocks is about 1,000 parts per million; if society eventually had to depend on average phosphorus, the costs of agriculture might become intolerably high. However, high-grade resources of phosphorus are very large: the present resource-to-demand ratio is 500 years for world reserves and an additional 800 years for potential resources. In addition, speculative resources are regarded as very large (but considerably smaller than fixed nitrogen from air or

TABLE 6. Average mole percent and composition of salts from seawater and of the gases in the atmosphere.

Seawater (except for water)					
$Cl_{0.4859}$	$Na_{0.4180}$	$Mg_{0.0485}$	$S_{0.0255}$	$Ca_{0.0091}$	$K_{0.0088}$
$C_{0.0021}$	$Br_{0.0007}$	$P_{0.000002}$	$Si_{0.0001}$	$(Fe,Al,Ti)_{0.0000005}$	
$Other_{0.0009}$					
Air (excluding variable amounts of CO_2 and H_2O)					
$N_{0.7805}$	$O_{0.2100}$	$Ar_{0.0093}$	$Ne_{0.0002}$	$(He,Kr,Xe)_{\ll 0.0001}$	

potassium from seawater). Nevertheless, phosphorus can hardly be regarded as inexhaustible. This led Wells *et al.*[9] many years ago to imply that ultimately we shall have to recycle bones as fertilizer.

With regard to the trace elements that are present in soil and are needed only at low concentrations, modern agriculture slowly depletes these elements. In the near term, shortages can undoubtedly be supplied from inorganic sources: in the long run, we shall undoubtedly be forced to return agricultural and animal wastes to the soil, particularly for the trace elements with limited resources, such as copper, zinc, and cobalt.

Beyond these nonsubstitutable elements, we use many rather scarce metals because of their special properties: copper because it is a good electrical conductor; nickel, chromium, and niobium because they confer corrosion resistance and high-temperature strength on iron; and mercury because it is a metallic liquid at room temperature. Goeller has studied in some detail the extent to which one might substitute for cadmium,[10] zinc, lead, copper, tin, and mercury.[11] His main conclusion is that, for most of their uses, substitutes derived from inexhaustible or nearly inexhaustible materials are available.

Would the aggregate additional cost (in both energy and dollars) of the substitutes or of the materials themselves be intolerable? To be more specific, let us estimate how much the aggregate unit cost of these metals could rise without causing the cost of avalloy to rise more than twofold.

We can look at the matter by again examining the chemical composition of avalloy. More than 95 percent of avalloy consists of iron, aluminum, silicon, magnesium, and titanium. These account for 95

TABLE 7. Present or future near-infinite resources for the most extensively used elements.

Element	Resource	Maximum concentration in best resource (%)	World resource (tons)	R/D ratio* (years)
CH_x (extractable)	Coal, oil, gas	<75	10^{13}	2,500
C (oxidized)	Limestone	12	2×10^{15}	4×10^{6}
Si	Sand, sandstone	45	1.2×10^{16}	5×10^{6}
Ca	Limestone	40	5×10^{15}	4×10^{6}
H	Water	11	1.7×10^{17}	$\sim 10^{10}$
Fe	Basalt,† laterite	10	1.8×10^{15}	4.5×10^{6}
N	Air	80	4.5×10^{15}	1×10^{8}
Na	Rock salt, seawater	39	1.6×10^{16}	3×10^{8}
O	Air	20	1.1×10^{15}	3.5×10^{7}
S	Gypsum, seawater	23	1.1×10^{15}	3×10^{7}
Cl	Rock salt, seawater	61	2.9×10^{16}	4×10^{8}
P	Phosphate rock	14	1.6×10^{10}	1,300
K	Sylvite, seawater	52	5.7×10^{14}	4×10^{7}
Al	Clay (kaolin)	21	1.7×10^{15}	2×10^{8}
Mg	Seawater	0.012	2×10^{15}	4×10^{8}
Mn	Seafloor nodules	30	1×10^{11}	13,000
Ar	Air	1	5×10^{13}	2×10^{8}
Br	Seawater	—	1×10^{14}	6×10^{8}
Ni	Peridotite	0.2	6×10^{11}	1.4×10^{6}

*R/D = resource to demand ratio in 1968.

†It must be noted that no process now exists for obtaining iron from basalt or nickel from the ultrabasic rock peridotite.

percent of the energy used to extract each ton of avalloy and 80 percent of the cost per ton. All these materials are in infinite supply, and their ultimate unit cost, in either energy or dollars, can hardly exceed today's unit cost by more than a factor of 2, as shown in Table 8. The remaining 5 percent of avalloy consists mainly of copper, zinc, manganese, chromium, lead, nickel, and tin; these represent about 20 percent of the total value of avalloy (Tables 3 and 4). At present, extraction from ore represents only about one-half the total cost of the latter metals. The aggregate cost of this group of metals that is sensitive to the grade of ore therefore is not more than 10 percent of the total unit price of avalloy. Thus, even if the price of these materials increased fivefold to tenfold, the unit price of avalloy would only double.

This is almost surely an overestimate, since we have not allowed for substitution. In 1968, for example, copper accounted for 12 percent of the total cost of avalloy in the United States. But copper is, in the long run, almost entirely replaceable by aluminum, titanium, and plastic. Moreover, manganese (in sea-floor manganese nodules) and nickel (in peridotites) are in nearly infinite supply; it is hard to see how the latter two metals can ever cost more than ten times their present price.

Some substitutes can be identified for the remaining elements—zinc, chromium, lead, and tin. Galvanized iron (which uses zinc) can, in good measure, be replaced by plastic-bonded steel; tin plate in cans can be largely replaced by plastic or glass. However, no very good substitute for chromium in stainless steel is presently known. Ultimately, this may force society to substitute titanium (which is in nearly infinite supply) for most uses that are now served by stainless steel. Nevertheless, the remaining elements—zinc, chromium, lead, and tin—could increase in cost by a very large factor, and the price of avalloy would still remain within a factor of 2 of its present real cost. This is true also of the remaining nonmetals, the metal oxides, the refractory metals, and the nonferrous by-product metals. Thus, we arrive at our basic observation: avalloy and demandite (with the extremely important exception of CH_x) are so heavily dominated by elements for which substitutes are available that their unit price is relatively insensitive to depletion of mineral resources. The tentative conclusion to be drawn is that, in principle, our social and economic structures are unlikely, in the long run, to be disrupted because we shall have to exploit lower-grade mineral resources—provided always that we find an adequate inexhaustible source of relatively cheap energy to substitute for CH_x. We put this hypothesis forward, realizing that it is in sharp opposition to the currently fashionable neo-Malthusianism; but we believe the neo-Malthusians have been misled by their habit of lumping all resources together without regard to their importance, ultimate abundance, or substitutability.

Mine Wastes

The catastrophists often point out that, as we exhaust our higher-grade resources and mine lower-grade rock, mine wastes will lead to disaster. But our analysis points to a rather different outcome. Since CH_x is by

TABLE 8. Energy requirements for the production of abundant metals and copper.

Metal	Source	Gross energy* (kWh/ton of metal)	E_L/E_H†
Magnesium ingot	Seawater	100,000	1
Aluminum ingot	Bauxite	56,000	1
	Clay	72,600	1.28
Raw steel	Magnetic taconites	10,100	1
	Iron laterites	11,900	1.17 (with carbon) ~2 (with electrolytic hydrogen)
Titanium ingot	Rutile	138,900	1
	Ilmenite	164,700	1.18
	Titanium-rich soils	227,000	1.63
Refined copper	Porphyry ore, 1% Cu	14,000	1
	Porphyry ore, 0.3% Cu	27,300	1.95

*At 40 percent thermal efficiency for generation of electricity.
†E_L/E_H=energy to extract from low grade ore/energy to extract from high grade ore

far the largest component of demandite, most of the waste from mining is associated with extraction of coal—8 tons of spoil per ton of coal mined in the United States. Thus, in the Age of Substitutability, when we no longer mine coal, the mine spoil per person associated with other sources of energy (breeders, fusion, and solar) will be much less than the mine spoil now associated with our energy system.[12]

As for the mine wastes per ton of avalloy, these can never increase much more than threefold because avalloy is dominated by iron, and high-grade taconite iron ore is only three times as rich as inexhaustible laterite ore. Moreover, the total waste from avalloy mining is only 5 percent of that from coal mining: a threefold increase in avalloy mining waste would still be small compared with the waste from coal mining.

We have not attempted to estimate other pollutants in the Age of Substitutability. However, since the bulk of air pollution (CO_2, CO, SO_2, trace elements, and fly ash) is the result of burning fossil fuels, there should be less air pollution in this asymptotic state than at present.

Although our estimates are reassuring, on the average, we are reluctant to leave an impression of facile optimism. The environmental impacts in specific places and specific situations might well be more serious than we have implied.

The Energy Budget

As we exhaust the high-grade materials and have to use lower-grade ones, the energy required to recover our needed materials will grow. Yet, because the composition of avalloy is so dominated by essentially inexhaustible iron and aluminum, the energy required to extract a ton of metal will not grow nearly as much as one might think. Although Frasché was correct in saying that "the extraction of mineral raw materials from low-grade rock is a problem in the application of energy—at a price" (Ref. 6, page 18), he should have added that the total mass of useful minerals that have a finite resource base is small. Therefore, the effect of their depletion on the entire energy system is less than Frasché's statement might imply.

Of course, even the elements in nearly infinite supply will take more energy to extract as we go from ores to more common rocks. The energy required for producing and recycling the abundant metals (magnesium, aluminum, iron, titanium) and copper were estimated by Bravard *et al.*[13] and are summarized in Table 8. In this table, we compare E_L, the amount of energy required to extract a ton of metal from low-grade, essentially inexhaustible ores (except for copper ore), with E_H, the energy required to extract the metal from high-grade ores.

It is remarkable that the estimated energy required to produce these metals from essentially inexhaustible sources is in every case not more than about 60 percent higher than the energy required to win the metals from high-grade ores. Even when all the reduced carbon is gone, the ratio for iron—by far the most important metal—is $\lesssim 2.0$.[14]

Extraction of useful metal from ore involves two separate steps: (1) mining and beneficiation of the ore and (2) reduction and refining of the metal from the beneficiated ore. Generally, the second step requires considerably more energy and expense than the first; for example, to mine and beneficiate 1 ton of iron from presently used magnetic taconite ores requires about 5 percent as much energy as is required for the total production of steel from ore. Thus, the overall energy required is not very sensitive to the grade of ore until the ore becomes extremely dilute, and this will never happen for the inexhaustible metals.

In 1968, the metal industry consumed 8.5 percent of the total energy used in the United States; some 90 percent of this was expended in the production of iron, aluminum, magnesium, titanium, and copper. The per capita annual energy expenditure for metal production came to about 8,000 kilowatt-hours per person, or 0.91 kilowatt per person. However, when only new metal from ores is considered, the primary metal industry accounts for 5.7 percent of total U.S. energy consumption; in this case, energy use is reduced to 5,300 kilowatt hours per person. In the Age of Substitutability (assuming electrolytic hydrogen, rather than carbon, is used), it seems fair to double the energy expended for these metals—to 10,600 kilowatt-hours, or a rate of 1.2 kilowatts per person—assuming that the amount of metal used per person and the composition of avalloy remain as they now are. Even in the Age of Substitutability, the amount of energy required per unit of metal to provide avalloy is hardly twice (rather than 10 or 100 times) that used in the present era. This is because the dominant metals—iron, aluminum, and magnesium—can be extracted from essentially inexhaustible resources, which demand relatively small additional amounts of energy for their extraction.

In our original paper, Goeller and I tried to estimate a per capita energy budget in the Age of Substitutability. Our main conclusion was that the use of lower-grade ores, and the possible requirement for conducting some agriculture with desalted seawater, would increase the energy demand per capita in the industrialized world by at least 2 kilowatt-years of heat per year per person (denoted simply by 2 kilowatts per person), and more likely by 4 kilowatts per person. This represents a relatively small increase for the United States—from 11 to 15 kilowatts per person; but for the world at large, which now uses 2 kilowatts per person, this increase would bring the asymptotic world demand to, say 5 kilowatts per person, and possibly even 7.5 kilowatts per person. Should the world's population grow to 10 billion, the world would expend energy at the rate of 50 and 75 terawatts—for simplicity, let us say 60 terawatts—indefinitely in the Age of Substitutability. This is eight times the present world energy budget of 7.5 terawatts and is one two-thousandth of the total energy the Earth absorbs from the Sun.

Can the World Reach and Sustain a 60-Terawatt Energy Budget?

Much of the neo-Malthusian literature begins with an implicit denial of the possibility of reaching and sustaining a 60-terawatt energy budget. Thus the Meadowses and Amory Lovins, at a recent workshop at the International Institute for Applied Systems Analysis, suggest that a more plausible future would consist of, say, 6 billion people each expending only 3 kilowatts per person—a total of 18 terawatts. The neo-Malthusians may prove to be right. Nevertheless, I believe it is important to examine discernible limits in an energy system that produces 60 terawatts. As we shall see, problems that are unimportant when the whole system is small become dominant when the system is extremely large. I believe it is useful to examine ultimate limits if for no other reason than to guide us in applying remedies now that would avoid foreclosure of a 60-terawatt world.

The Exhaustion of Fossil Fuel

The total amount of CH_x recoverable in principle is estimated at about 10 trillion tons of coal equivalent: this has an energy content of 2×10^5 quads (1 quad $= 1.05 \times 10^{15}$ kilojoules), or about 6700 terawatt-years. At 60 terawatts, the fossil fuels would last little over 100 years. Since it is most unlikely that all of the fossil fuels can be recovered, the actual time for exhaustion would be considerably less than 100 years, once the asymptotic level of 60 terawatts were reached.

The CO_2 Catastrophe[15]

The use of fossil fuels may have to be ended much more quickly than this, even if we never reach 60 terawatts, because of the possible warming of the Earth's surface resulting from release of carbon dioxide (CO_2), the so-called "greenhouse" effect. The atmosphere contains about $2{,}300 \times 10^9$ tons of CO_2 corresponding to a concentration of 330 parts per million. The amount of CO_2 in the atmosphere has been increasing at a rate of about 10×10^9 tons per year since 1958, when C. D. Keeling began accurate measurements of atmospheric CO_2. This equals half of the CO_2 thrown into the atmosphere by the burning of fossil fuel.

Manabe and Weatherald estimate that a doubling of CO_2—that is, from $2{,}300 \times 10^9$ tons to $4{,}600 \times 10^9$ tons—might raise the *average* temperature of the Earth's surface by 2.5 °C, and the polar temperatures by 8 °C. So drastic an increase in the Earth's surface temperature would be almost without precedent, and would surely have profound social consequences.

If 50 percent of the CO_2 produced by burning fossil fuels continues to remain in the atmosphere even as the total CO_2 burden increases, then to double the CO_2 concentration would entail the burning of a little more than 1 trillion tons of coal—only about 10 percent of the estimated theoretically recoverable resource. Even at the present rate of 7.5 terawatts, if all our energy were derived from coal, 1 trillion tons would be exhausted in about 100 years.

There are great uncertainties in this, and I mention the two most puzzling. First, at the same time we are burning vast amounts of fossil fuel, we are also clearing the tropical forests. It is rather unclear how much of the observed increase in CO_2 comes from forest clearing—indeed, whether the biosphere at present is a net source or a net sink of CO_2. Second, even the elaborate global climatic model of Manabe and Weatherald is oversimplified. For example, it does not include the effect of the increased cloudiness that would ensue were the oceans to warm even slightly. Thus the prediction of the effect of increased CO_2 is itself ridden with uncertainty. What we can say with certainty is that CO_2 in the atmosphere has been increasing during the 20 years that D. Keeling has been measuring it accurately; that the preponderance of scientific opinion holds that this increase is associated with the burning of fossil fuels; and that this could lead to catastrophic climate change. Carbon dioxide must be viewed as an ominous, though uncertain, Sword of Damocles that dangles over our industrial society.

Effect of Direct Heating

The heat directly thrown into the atmosphere if energy is used at the rate of 60 terawatts continuously would seem to affect the climate less than does the CO_2. If the Earth is viewed as a black body, we can easily compute that 60 terawatts would increase the radiative temperature of the Earth by about 0.03 °C. This in itself seems a reassuringly low figure, especially when compared with the estimated 2.5° increase in surface temperature caused by a doubling of CO_2—which would possibly result if 60 terawatts of heat were supplied by coal for about 10 years. One must remember, however, that the localized heat pollution could be severe. It has been estimated[16] that, if the heat load exceeds 1 megawatt per hectare and the total load is of the order of 10,000 megawatts, local turbulence and even tornadoes might be created. Thus, even if the global effect of 60 terawatts of energy were tolerable, its effect on local climate may be considerable.

The Alternatives: Fusion, the Sun, the Breeder, Geothermal

When the fossil fuels are gone, we will have to base our energy system on some combination of geothermal, fusion, solar, and the fission breeder. Of these, we have in hand geothermal energy, but it is not clear how large this resource is. Because the total steady-state geothermal flux over the continents is about 8 terawatts—about what society now uses—it is hard to imagine geothermal ever contributing a large fraction of the 60 terawatts. As for fusion, although progress continues to be made, we cannot count on its technological or economic feasibility. Thus, to be realistic, I would say we shall probably have to depend primarily on some combination of the Sun and the breeder reactor: both are technologically feasible, both are essentially inexhaustible resources (the breeder is so because it can burn the low-grade uranium and thorium from granitic rocks or from the sea). To most neo-Malthusians, the Sun seems to be the preferable long-term energy source, despite its intermittency and probable high cost. Nevertheless, it seems difficult to envisage a *high*-energy society—say, one using 60 terawatts—based on solar energy alone. If we want a high-energy society, I would judge the odds are high that it would have to be based on breeder reactors.

Can We Live with Fission?

We must therefore confront an underlying question: Can we live with fission, especially if it provides a large fraction of the 60-terawatt budget? By now, the debate over nuclear energy has taken us very far from the simplistic utopia I imagined ten years ago. At that time, I considered it plausible that fission based on breeders would be a cheap as well as an inexhaustible energy source. Unfortunately, my hopes about the cost of fission energy have not been borne out—whereas fixed-price contracts for nuclear plants were being let for about $100 per kilowatt in 1967, the price today is at least ten times higher in current dollars, perhaps five times higher in 1967 dollars. Thus the dream of cheap energy from breeders certainly has dimmed; on the other hand, the estimated cost of energy from solar sources is also much higher, and nuclear energy continues to look a good deal cheaper than most forms of solar energy (hydroelectric and wind power are exceptions to this).

The problems of nuclear energy take on new dimensions if one assumes that 60 terawatts of energy will be produced by breeders. Let us assume that each breeder produces 6000 MW of heat, and about 2000 megawatts of electricity; this is about twice as large as the largest nuclear reactors now in operation.

The entire global system, under these assumptions, would comprise 10,000 breeders containing 100,000 tons of plutonium-239, of which at least 20,000 tons would be processed each year.

One could go on with such statistics—say, the amount of plutonium-239 processed per day, or the number of shipments of radioactive fuel, or the amount of waste produced per year—all would be very large. But I think the point is unmistakable: to support a 60-terawatt society with fission breeders would require an immense nuclear system, so large that one must ask whether the system is credible.

What really is the difficulty? Basically, it is that, because failure of the system is a matter of probability, the failure rate increases as the system grows, and it becomes unacceptably large if the system is large enough. The Rasmussen study[17] estimated that the probability of a meltdown in a light-water reactor that would release kilocuries or more of radioactivity is around 1 in 20,000 per reactor per year, the uncertainty being a factor of 10 in either direction. If the same probability held for all 10,000 breeders (in itself a large assumption), then the failure rate becomes about 1 every 2 years. Most of these failures would be innocuous, but, unless the world's attitude toward radiation hazard changes, I would judge a meltdown accident every other year would not be acceptable.

The same sort of argument can be made for each of the other components of the system and its "failure modes": proliferation, clandestine diversion, accident during transport, safe handling of wastes. The *a priori* probabilities are small when the system is small, and they become larger as the system grows larger—assuming, of course, that the technologies do not improve.

This, I believe, is the main weakness in the foregoing speculation: the technology of nuclear safety will almost surely improve, as will the technology of preventing clandestine diversion. Both matters are under intense study, and there is little reason to doubt that "technological fixes" for both these problems will be forthcoming. Indeed, I would say the nuclear system, just as the air transport system, can expand only to a point where its failure rate is considered tolerable. Air travel would have been impossible today had airplanes been as subject to failure now as they were 40 years ago. One would hope—indeed, it is a requirement—that, as the nuclear system expands, its failure probability per reactor diminishes in proportion.

Nevertheless, additional social as well as technological fixes will be needed if we are to contemplate an all-fission, 60-terawatt world. Two social fixes strike me as being particularly important: first, the creation of a corps of nuclear experts, along with the traditions that go with such a corps, not unlike airplane pilots and their traditions; and, second, the investment of the system with a kind of social stability and permanence.

The importance of social stability was illustrated by the incident at Da Lat involving a small research reactor operated by the South Vietnamese. Rather than allow the reactor to fall into North Vietnamese hands, South Vietnam unloaded the fuel, shipped it to the United States, and demolished the unloaded reactor. The maneuver was successful, but would it have worked were the reactor a 6×10^6 kilowatt (heat) breeder rather than a research reactor only a thousandth the size? Large reactors require active surveillance and care, even for some time after they are shut down. Their presence would tend to demand of the society a kind of stability that is hardly required by other technologies of energy generation.

One can be either a pessimist or an optimist as one contemplates this situation. The pessimists would argue that such stability is unprecedented and cannot be guaranteed—therefore, nuclear fission is unacceptable. The optimists, among whom I count myself, would argue that the *presence* of nuclear reactors would tend to impose limits on the degree of violence that would accompany social change. To take the Vietnam example again, had the Da Lat reactor been a very large breeder, it obviously would have been in the interest of both sides to avoid an accident that might contaminate the land; an orderly takeover, much as with an ordinary power plant, would have been the rational action.

It is concerns of this nature that have led me to refer to nuclear energy as a Faustian bargain: the society, in exchange for an inexhaustible energy source, must exercise a degree of technological prudence as well as maintain a social stability that goes beyond what we have hitherto required. Whether we are up to this challenge—whether we can, in the long run, live with fission—is a question that none of us can answer today.

The Transition

But all this is far in the future: the Age of Substitutability is a mirage, unless we can move from where we are now to this asymptotic world without destroying the social edifice. Several interrelated issues emerge. First, it is reassuring that the neo-Malthusian claim that we will run out of mineral resources, or that the energy demand or the pollution load per unit of mineral will skyrocket, seems to be without merit—this I believe is clear from Goeller's work. But whether our present market mechanisms are sufficient to achieve an orderly transition is something else. To take an example, aluminum from the all-but-inexhaustible kaolinite clays might cost 50 percent more than aluminum from bauxite. There is little doubt that a transition to kaolin will eventually take place. But, as was the case when the price of oil more than doubled, in such a transition it is the poorer countries that suffer more than the rich. Can the poor countries be helped sufficiently during this transition to avoid collapse?

Second, our analysis once more reinforces the centrality of energy in the Age of Substitutability. Although the energy per unit product is not likely to increase by more than a factor of 2, if that, our assumption of a world of 10 billion at an energy demand per capita of 6 kilowatts leads to a total energy demand of 60 terawatts—and this, to my mind, poses difficulties no matter how we propose to meet such a demand.

The prime importance of conservation is evident: if the energy budget could be reduced to what the neo-Malthusians propose—say, around 20 terawatts (that is, 6 billion people at 3 kilowatts, rather than 10 billion at 6 kilowatts), the energy system, whether nuclear or nonnuclear, would be smaller and its systems problems would be lessened. But these are only hopes—partly because an asymptotic world with a population of 6 billion can hardly be assured, partly because the underdeveloped countries will continue to aspire to a standard of living much closer to that of the developed countries. A world of rich and poor living contentedly side by side simply does not seem very likely—instead, Heilbroner's "wars of redistribution,"[18] pitting the overpopulated have-nots against the underpopulated haves, could collapse the society long before our carbon runs out. And, despite the most stringent conservation, I should think that an ultimate world expenditure per capita of 5 or 6 kilowatts is not excessive, considering that, in the Age of Substitutability, the energy per unit of avalloy may increase by about a factor of 2.

It would seem to me that, in view of the uncertainties, we ought to be planning, even now, on ways of creating a livable 60-terawatt world, even a world largely based on fission. For it is likely that, just as we learned how to reduce airplane failures to manageable proportions, so we could expect our fission system, as it expands, to improve. Ideas about how to cope with a very large fission system have been put forward by the Institute for Energy Analysis[19] and the International Institute for Applied Systems Analysis.[20] In general, these ideas center on the confinement of the nuclear enterprise to as few sites as possible. The sites themselves would be rather isolated and self-contained: this would minimize transport of radioactive material as well as reducing the amount of land that could conceivably be contaminated by an inadvertent release of radioactivity. Beyond these technological advantages, dedication of such sites to nuclear activities would be expected to invest the nuclear enterprise with a sort of permanence and long-lived institutional infrastructure, in much the same way that the dike system of Holland or the national park system in the United States have acquired permanent infrastructures.

To be sure, many of these ideas are just beginning to emerge, and they evoke much controversy. The institutional demands of a long-term, very large nuclear system are just beginning to be explored, and to review this growing literature here would take me too far from the Age of Substitutability. Suffice it to say that the character of the Age of Substitutability—whether it is brutish and Malthusian or reasonably comfortable and autarkic—may well depend on how well, in the very long run, we can live with fission.

References

[1] D. H. Meadows, D. L. Meadows, J. Randers, W. W. Behrens III, *The Limits to Growth* (Universe Books, New York, 1972).

[2] M. Mesarovic and E. Pestel, *Mankind at the Turning Point* (Dutton, New York, 1974).

[3] K. E. Boulding, *The Meaning of the Twentieth Century: The Great Transition* (Harper & Row, New York, 1964).

[4] N. Georgescu-Roegen, *Entropy Law and the Economic Process* (Harvard University Press, Cambridge, MA, 1971).

[5] H. Barnett, "The Myth of our Vanishing Resources," Trans-Action **4**, 6–20 (1967).

[6] D. F. Frasché, "Mineral Resources," a report to the Committee on Natural Resources, National Academy of Sciences, Publication 1000-C, December 1962, p. 18.

[7] (See Table 1.)

[8] H. E. Goeller, paper presented at the University of Minnesota, Forum on Scarcity and Growth, sponsored by the National Commission on Materials Policy, Bloomington, MN, 1972.

[9] H. G. Wells, J. S. Huxley, and G. P. Wells, *The Science of Life* (Doubleday, Garden City, NY, 1931), Vol. 3, pp. 1031, 1032.

[10] W. Fulkerson and H. E. Goeller, eds., "Cadmium, the Dissipated Element" (ORNL) NSF-EP-21, Oak Ridge National Laboratory, Oak Ridge, Tennessee, 1973.

[11] H. E. Goeller, paper prepared for the *ad hoc* committee on the rational use of potentially scarce metals, Scientific Affairs Division, North Atlantic Treaty Organization, London, 1975.

[12] For example, see A. M. Weinberg and R. P. Hammond, American Scientist **58** (4), 414 (1970).

[13] J. C. Bravard, H. B. Flora II, and C. Portal, "Energy Expenditures Associated with the Production and Recycle of Metals," ORNL-NSF-EP-24, 1972.

[14] Could we imagine charcoal from wood, used to reduce iron ore until 150 years ago, replacing coke today? Is this practical?

[15] C. F. Baes *et al.*, "Carbon Dioxide and the Climate: The Uncontrolled Experiment," American Scientist **65**, 310–320 (1977).

[16] United States Nuclear Regulatory Commission, "Nuclear Energy Center Site Study—1975," Executive Summary, NUREG-0001-ES, 9, Springfield, VA, National Technical Information Service, 1976.

[17] United States Nuclear Regulatory Commission, "Reactor Safety Study: An Assessment of Accident Risks in U.S. Commercial Power Plants," WASH-1400 (NUREG-75/014) (Washington, DC, 1975).

[18] R. L. Heilbroner, *An Inquiry Into the Human Prospect* (Norton, New York, 1974).

[19] A. M. Weinberg, "Outline for an Acceptable Nuclear Future," ORAU/IEA (0) 77–17 (Institute for Energy Analysis, Oak Ridge Associated Universities, Oak Ridge, TN, 1977).

[20] W. Haefele and W. Sassin, "Contrasting Views of the Future and Their Influence on our Technological Horizons for Energy," pp. 196–225, in *Future Strategies for Energy Development—A Question of Scale*, ORAU-130, proceedings of a conference at Oak Ridge, Tennessee, 1976, sponsored by Oak Ridge Associated Universities.

Energy Wars

There is a growing polarization as to the course of development of our energy system. On the one side are the energy "radicals," such as Amory Lovins, Barry Commoner, Ralph Nader. For them, Lovins's Soft Energy Path[1]—meaning a small, decentralized, largely solar-based system—is a necessity. Hard technologies, particularly nuclear energy, are unclean, unsafe, unnecessary. On the other side are the energy "conservatives"—the Energy Establishment and the nuclear people. For them the Soft Path is impossibly idealistic, impractical, and elitist. The present course, largely dependent on nuclear energy and other "hard" technologies, is both necessary and manageable.

The debate over solar energy and nuclear energy, over centralized and decentralized energy systems, over electrical and nonelectrical energy, is becoming more strident. To be sure, actual fighting has broken out only over the nuclear issue: the fatality at Super-Phenix, the demonstrations at Seabrook and Kalkar, the sit-ins at Whyl. But the nuclear issue is only the most acute manifestation of an underlying confrontation between those who believe our present energy system is fundamentally on the right track and those who insist that we must switch directions. Should the polarization harden and acquire the aspects of a religious war, one can foresee strife breaking out over big coal-powered plants, coal-hydrogenation plants, and the like, as well as nuclear plants.

Can the skirmishes be headed off before they develop into a full-fledged war? Can we not conceive of some middle ground that includes

From American Scientist **66**(2), pp. 153–158 (1978). Reprinted by permission.

both solar energy and nuclear energy and that both sides could agree was better than all of one or all of the other?

Thermodynamic Imperatives

For many energy radicals, our energy system is basically wrong because it ignores thermodynamic imperatives. The most obvious of these is the so-called first-law efficiency. This is defined simply as the work done in a process divided by the heat energy or enthalpy put into the process, or, more generally, the enthalpy usefully used divided by the enthalpy put into the system. An automobile that gets 30 miles per gallon is more efficient and energy saving than one that gets 20 miles per gallon; a house heating unit that requires 150,000 Btu per hour of fuel burned to deliver 100,000 Btu per hour of heat is more efficient than one that delivers only 50,000 Btu per hour. If energy is a scarce resource, then clearly there is advantage in using it as efficiently as possible, and radicals and conservatives agree that high first-law efficiency is desirable.

But this in a way is misleading. As the late Joseph Keenan pointed out, there can be no scarcity of energy because, according to the first law, energy is conserved. Rather, what we lack is energy in a useful form. The usual thermodynamic measure of the usefulness of energy is the *availability*, defined by Willard Gibbs as equal to $E + P_0V - T_0S$. Here P_0 and T_0 are the pressure and temperature of the ambient environment to which heat is to be rejected; E, V, and S are, respectively, the internal energy, volume, and entropy of the system. (Note that the availability and Gibbs free energy become identical when the environment and the system are in equilibrium.) What we call an energy crisis really ought to be called an available-energy (or even free-energy) crisis: it is these that must be husbanded and that are not everywhere free for the asking. Thus, the availability of energy from highly superheated steam is significantly greater than that from a larger amount of saturated steam, even if the enthalpies are the same in the two cases. On the other hand, the availability of energy in a piece of coal or other fossil fuel is almost identical with its enthalpy or even its internal energy; it is for this reason that, in ordinary parlance, availability and energy are spoken of interchangeably.

Since energy at high availability is relatively scarce, whereas energy at low availability is abundant, it makes thermodynamic sense to use energy of low availability for tasks that can be done with low availability, energy of high availability only for tasks that really require high availability. Electric resistive heating hurts the sensibilities of the energy radicals because one is using energy at extremely high availability (electricity) simply to raise temperatures a few degrees above ambient. This is inelegant and wasteful. To a degree, radicals and conservatives agree that conserving availability is desirable: one ought to match the energy source to its end use, to maximize, as the current jargon has it, the second-law efficiency as well as the first-law efficiency.

Beyond the Thermodynamic Imperatives

What then is the basis for the energy war, since both sides agree that our energy system ought to strive to maximize first- and second-law efficiencies? As Rotty and Van Artsdalen point out,[2] there are imperatives, other than thermodynamic imperatives, that often conflict with thermodynamic imperatives: environment, economics, reliability, time—in short, the texture of our social fabric. Radicals value many of these imperatives quite differently than do conservatives. For example, although conservation of energy is in principle a good thing, its achievement may be too expensive in practice, or it may reduce the reliability of the system. Conservatives are less ready than radicals to sacrifice these other goals in order to save energy.

In designing an energy system, no less than an energy-transforming device, there are many factors, other than those that derive from thermodynamics, that must be considered. This is almost too obvious to deserve emphasis. I stress it because in some of the radical literature, notably Barry Commoner's *Poverty of Power*, and even Amory Lovins's *Soft Energy Paths*, thermodynamic imperatives are elevated to sacred dogma. First and foremost, the system must conform to these imperatives. Economics, reliability, time—these seem to carry much less weight.

Obviously, we must conserve energy, not only because our sources of high-availability energy—fossil fuels or nuclear fuels in burner reactors—are limited, but because energy transactions pollute the environment. But, except for the doctrinaire conservationists, perhaps because they are opposed to growth, conservation *per se* is not a transcendent purpose of our society. Should we be successful in developing inexhaustible energy sources, our underlying rationale for conserving energy would shift from the present first-order concern over dwindling fuel resources to less tangible issues: subtle environmental effects, cap-

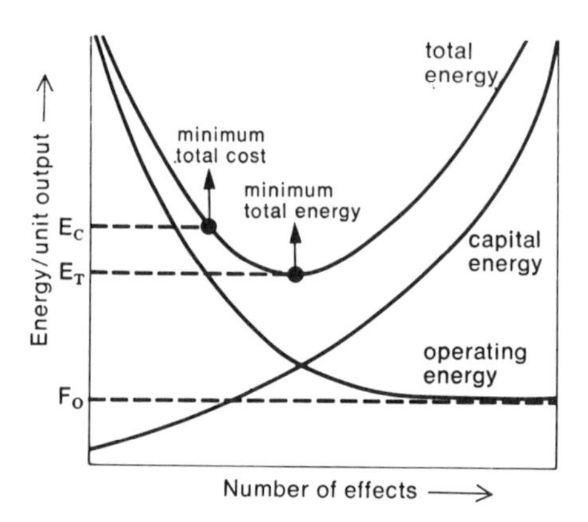

FIGURE 1. *Total energy required to desalt seawater is the sum of the operating energy and the energy that is embodied in the apparatus (capital energy). Note that the minimum total energy is larger than the thermodynamic minimum (F_0), which is equal to the change in free energy.*

ital expenditures, or even social risks, such as proliferation, that some inexhaustible energy sources may entail.

Generally, energy conservation exacts other costs. Consider the desalting of seawater by multiple-effect distillation. The process can be performed more and more reversibly by increasing the number of effects—that is, by reducing the temperature differences between effects. Ultimately, if the number of effects becomes extremely large, the process energy approaches the thermodynamic minimum, 3 kilowatt-hours per 1,000 gallons, which is the difference in free energy between the seawater and the distilled water. If we plot the energy required to operate the plant as a function of the number of effects, daily output being held constant, we obtain a monotonic decreasing curve that approaches an asymptote, F_0, which is the aforementioned difference in free energy between the two states (Figure 1). But to approach this minimum requires us to build a very elaborate plant: in effect, to exchange capital (dollars) for operating energy.

Actually, the minimum energy that must be used is, even theoretically, larger than F_0 because we have not included the energy required to build the plant. This "capital" energy dominates when the number of effects becomes very large. The energy expended to conduct the process is the sum of operating energy and prorated capital energy, and this total energy must go through a minimum, E_T, at some value that exceeds the thermodynamic limit F_0.

In practice, plants are designed to minimize the total monetary cost of the product. On the curve of total energy cost vs. plant size, the actual design point corresponds to an energy expenditure E_C; the theoretical maximum energy that can be saved is, as pointed out by van Gool, $E_C - E_T$, and this is less than $E_C - F_0$.[3] To achieve this saving requires expenditure of money, but if too much money is spent (to make the plant even more elaborate) the energy expended increases rather than decreases!

There is an even more fundamental trade-off involved in saving energy, and that is time (as was pointed out by D. Spreng[4]). If we ask why we have, ever since the beginning, chosen to use more energy, it is probably because we value time more than energy. Remember that thermodynamics demands immortality of us if we are to conduct our lives with minimum expenditure of energy: reversible processes require either infinite time or infinite capital energy. But we are mortal: we know, each of us, that our time on Earth is limited, and at least in Northern European and American cultures (which do not view an eternal afterlife as being a simple extension of our time on this globe), I suspect time is viewed as one of our most precious resources. By expending energy, we can save time: there is a sort of dichotomy between energy and time that primitive people sensed when they domesticated animals, and then, even better, acquired intelligent slaves.

Still another characteristic that transcends thermodynamics is reliability. Here I speak of reliability in the sense of equipment reliability and system resilience. Some systems are inherently more reliable suppliers of energy than others—either because the devices are mechanically more robust, or because the fuel system is more certain. Solar energy is intermittent; the diurnal cycle is superimposed on a somewhat random cloudiness. If reliability is valued highly (and I think this valuation may be related to our perception of time), then an acceptable solar energy system would have to be backed up with a large storage system. If the system were to be completely reliable, it would, in principle, have to have an extremely large storage system, and this would be expensive: one buys reliability of energy supply by raising the cost of a delivered unit of energy.

On a large scale, one must consider the reliability of an entire system—what the ecologist C. S. Hollings calls the resilience of the system[5]. By resilience, Hollings means the capacity of a system to absorb external perturbations without collapsing. Typically, a centralized system—for example, a centralized electrical grid—is less resilient, more vulnerable than a decentralized system. This is one of the main thrusts of the argument against centralized electrical systems: they are vulnerable to collapse, as we have seen in the Northeast blackouts—a potent argument against the centralized systems. Still, by spending money, one can increase resilience: indeed, it is believed that the Western European electrical grid is much more robust than the

U.S. system because the interties are so much stronger than those in the United States, especially in the Northeast. If one were to contemplate an energy system based *entirely* on the Sun, the entire system would also lose resilience: it would be subject to a common mode failure—a protracted period of cloudiness. To protect against this contingency, large expenditures would be required to provide energy backup from a grid or storage system.

The Nuclear/Solar Debate

The preceding, somewhat abstract, discussion was intended to press home the primary point: thermodynamic considerations alone cannot be a solid basis for establishing energy policy. There are a host of other complex and interrelated factors—environment, economics, time, reliability, resilience—that demand consideration. To focus too strongly on the thermodynamic imperatives tends to obscure the real problems of formulating a proper energy policy.

The matter has come to a head in the increasingly violent nuclear debate. Stripped of excess rhetoric, the debate hangs on the basic question, "Can the Sun replace uranium?" I pose the question this way rather than "Can we do without uranium?" because, ultimately, unless fusion works, the other alternatives, fossil fuel and geothermal, will not do. Fossil fuels are limited in extent and are beset with the CO_2 uncertainty. The geothermal energy flux is limited; therefore, although the total geothermal energy reservoir is very large, it can be extracted only slowly: at the moment, the geothermal flux on the continents is about equal to man's energy flux. Only the Sun in its various manifestations [direct solar, hydro, wind, waves, ocean thermal energy conversion (OTEC), biomass], and fission based on breeder reactors are *technically* feasible as well as being essentially inexhaustible energy sources.

The Solar Paradigm

What, then, are the underlying tradeoffs between an energy system based primarily on the Sun and one based primarily on uranium breeders? I believe it is necessary to examine the implications of a full commitment to the one or the other. For the energy revolutionaries are implacable: it is not some combination of the Sun and uranium that they seek. Uranium is an abomination that must be extirpated. If this is so, then, in considering the advantages and disadvantages of a solar society, one must examine the consequences of a completely successful solar revolution: the Sun must stand by itself, except for a little help from geothermal energy.

In this context, one immediately encounters the essential weaknesses of the all-solar scenario: the Sun is intermittent and it is diffuse. If the all-solar society values time with anywhere near the same weight as I believe we do, then an energy system based on the Sun must come to terms with intermittency: it must approach the reliability of our present system. And if the social structure of the all-solar society is to resemble our present structure, we must fit a diffuse energy source into a concentrated, largely urban society.

Reliability will require storage, and storage is expensive. I know of no real analyses of what storage would cost in an all-solar energy system. If the system were based on photovoltaic arrays, storage would have to be electrical—say storage batteries. To take care of a six-day cloudy period, storage alone might cost as much as 70 cents per annual kilowatt-hour (akWh). The added costs for collectors and photovoltaic surfaces leads to a photovoltaic system that costs about $1.40/akWh. Storage cost in a system that stores heat rather than electricity—the power tower—is less; on the other hand, judging from the costs incurred in the construction of the 5 megawatt Sandia power tower, it seems to me unlikely that the total cost will be less for a power tower than for photovoltaics—around $1.40/akWh. This may be compared to the projected cost of the breeder—say $1,500/kW to $2,500/kW, or, at 70 percent capacity factor, 25¢/akWh to 42¢/akWh. Even if transmission costs and operating costs are added for the breeder system, and assuming that the photovoltaic system is dispersed and requires no transmission, it is hard to see how an electrical all-solar energy system that possesses the same degree of reliability as the alternative based on nuclear energy could be as inexpensive as the nuclear alternative. This is not to say that we ought not pay something extra for solar in order to avoid the problems of nuclear energy. Rather, at this stage, we simply cannot say what rejection of nuclear energy would cost nor what society can afford.

There are other possibilities. The ocean thermal system has its own storage system in the oceans; but it is highly centralized and requires a transmission system. Although several estimates have been made of its capital cost, I would be skeptical of any such estimates until an actual OTEC plant is built.

Because the Sun's energy is so diffuse, most visualizations of an all-solar society try to make a virtue of diffuseness. After all, almost half of our *uses* of energy, particularly for space and water heating and for industrial processes, are at relatively low temperatures ($<200\,^{\circ}C$) and are rather diffuse. Let us use the Sun for those jobs, thus largely avoiding the use of electricity, say the energy radicals. Electricity is viewed as being tainted, especially if it is generated centrally. In the radical view, such centralization of the energy system implies an authoritarian society.

The logic here is appealing, especially since solar heating of water and of space is a demonstrated technology, but what about the cost? At some price of electricity, solar heating is surely competitive, but it would be rash to say what that price will be. If heat pumps with a coefficient of performance of, say, 3.5 can be developed, then electricity delivered at 5¢/kWh would supply heat at about $\$4/10^6$ Btu—corresponding to oil at $24 per barrel. Can a solar-heated house meet this? Probably so, in many locations, if the system can be backed up with a fairly cheap alternative when the sun is not shining; probably not, in many locations, if the system must be 100 percent solar.

What about transportation in an all-solar society? Usually it will depend on biomass converted to alcohol, or, as M. Calvin suggests, on hydrocarbons produced directly from such plants as euphorbia. Again, the technology, at least for alcohol production, is available; the question always is, How much can we depend on biomass? Can we count on it for a large share of our energy system, say much of our transport sector? The problem, of course, is the commitment of land. At an overall photosynthetic efficiency of 0.6 percent—which may be compared with the average for the world's land plants of 0.35 percent—it would require about 10,000 square miles to grow biomass for an annual expenditure of 1 quad ($= 10^{15}$ Btu $= 1.05 \times 10^{15}$ kilojoules). To supply the present United States transport system would require about 18 quads per year—say 180,000 square miles devoted to energy farms. This is not impossible. But are other ways better?

Production of hydrogen by photovoltaics is a good deal more efficient. Consider a photocell with efficiency of 15 percent. Even if it intercepts but 30 percent of the Sun's rays and thus generates electricity with an overall efficiency of, say, 5 percent, this can be converted to hydrogen at an overall efficiency of 3 percent, some 5 times higher than the 0.6 percent we assumed for direct fixation with biomass. On the other hand, hydrogen produced by electrolysis is more expensive than biomass. At 5¢/kWh, one estimates that the hydrogen costs $20/MBtu or the equivalent of oil at $120/bbl. Most estimates of the cost of biomass are a good deal lower.

The transport system could, in principle, be electrified. Whether batteries can be developed that will make better economic sense than a system based on biomass is a question I do not believe anyone can now answer. Indeed, I do not believe it is possible to say whether the evolution of our energy system ought to progress along the electrical path or upon the nonelectrical path. To the argument that many of our end uses do not really require electricity, one can reply, but every one of our end uses *could* be done electrically. If electricity were cheap enough, and if its use entailed little dislocation, then for every solar utopia one could construct an electrical utopia that performs all of the end uses more reliably, more conveniently, and, in most cases, more cheaply.

The Nuclear Paradigm

Thus the electrical utopia is the mirror image of the solar utopia. Energy is generated in large central power plants; the society is largely electrical. Although this in some cases leads to lower second-law efficiencies, the society gains in convenience, reliability, and possibly cost.

The electrical utopia, no less than the solar utopia, is possible. From what we now know, it ultimately depends on the wide deployment of nuclear breeders or nuclear fusion. If, as we assume, fusion cannot be counted upon, then our future electrical utopia derives its energy from many thousands of nuclear breeders, of which about 1,000 would be in the United States. These would mainly produce electricity, which would be the backbone of the entire energy system.

The difficulties of the very large nuclear system have been argued at great length, and I shall not repeat them. They hang mainly around the twin risks of radioactive contamination and proliferation. Nevertheless, it must be conceded that a *properly* operating nuclear system—one that sequesters its wastes properly and is not used as a source of illicit bomb material—is a largely benign source of energy. Its shortcomings arise only if there is a failure in the system.

Although most of the current argument about nuclear energy hangs around the issues of waste disposal and proliferation, my own view is that these are lesser issues: waste disposal because, as the recent American Physical Society study has verified, the technical problems are largely resolved and await only a full demonstration[6]; proliferation because nuclear power is a sufficient but not a necessary condition for proliferation. Nuclear power plants might ease the path toward nuclear bombs for states that make the decision to go nuclear impulsively and in a short time; they hardly bear on states such as Taiwan, Israel, Korea, South Africa, Pakistan, Brazil, Argentina, which might decide to go nuclear as a deliberate, long-term national policy. To hang nuclear power with the onus of proliferation when most proliferation is likely to occur, and indeed has already occurred, via other pathways, is a bit like looking for one's watch not where it was lost but under the lamppost where the light is better.

I would say the primary, long-term concern over nuclear power is the possibility of a catastrophic accident. A 1,000 megawatt electric (MWe) nuclear reactor contains 15 billion curies of radioactivity; immediately after shutdown, some 200 megawatts of heat are generated by radioactive decay, and it is several weeks before the reactor can do without any external cooling. The probability of a meltdown accident that would spread a significant amount of radioactivity—say several kilocuries or more—is, according to the Rasmussen report, about 1 in 20,000 per reactor per year.[7] Most such accidents would do very little harm. They would be of the order of the Windscale incident in 1958, when about 30 kilocuries of iodine-131 were spread over the countryside. Nevertheless, unless the public perception of nuclear energy changes, I would expect that even such an incident as Windscale would now prove unacceptable. The aim of the nuclear enterprise, at least until public perceptions change, must be to avoid accidents. The public does not understand probabilities, especially very low probabilities; the public understands only no accident, no consequences.

Although a probability of 1/20,000 per reactor per year is low, it may not be low enough should the nuclear enterprise really expand. In our electrical utopia, there might be several thousand breeders, and, if the probability of meltdown per breeder per year is no better than the Rasmussen probability, then one might expect a meltdown every few years. The nuclear enterprise cannot expand beyond a point at which the accident rate is acceptable. Unless the accident rate is reduced considerably below the Rasmussen probability, I would doubt that the electrical utopia based on breeders is likely to come about.

What then must be done to make nuclear energy acceptable? Our Institute for Energy Analysis has been pondering this question for the past year; what I present is a brief outline of our findings. We seek a long-term nuclear energy system that is both acceptably safe and acceptably proliferation-resistant. These two requirements are not identical, and indeed they may work at cross-purposes in some respects. For example, if proliferation hardening requires much more shipment of radioactive fuels to a central reprocessing center, then the system is more, not less, vulnerable to accidents that might cause spread of radioactivity during transportation of the fuel.

The key to increasing the safety of the nuclear system is the nuclear energy center. Nuclear energy ought to be confined to as few places as possible—perhaps 100 centers altogether in the United States, plus a few waste-disposal sites—amounting in all to some 5,000 square miles of land committed, essentially in perpetuity, to the nuclear enterprise. Indeed, the most important single action that can now be taken toward achieving an acceptable nuclear system would be to confine all future reactors essentially to the existing 100-odd sites.

The advantage of such siting policy seems compelling to me. First, one thereby minimizes the land that can be placed at risk of radioactive contamination. Second, the cadre which would be in charge of the activities at such centers would be much stronger than that which would be available at more dispersed sites. In final analysis, the safety of the nuclear system depends on the strength, the professionalism, and the dedication of those in charge. This is surely enhanced by coalescing the activities into a few powerful centers that possess many of the attributes of Oak Ridge, Hanford, Savannah River, or Los Alamos. Third, the centers could afford much more stringent security than would be conveniently feasible in smaller centers. Transport of radioactivity and of plutonium would be largely internal; this would minimize the possibility of diversion. Moreover, one can envision a system of *resident* international inspection, as an extension of the International Atomic Energy Agency spot-inspection system, that could place added obstacles in the path of states intent on using their nuclear facilities to make bombs—for example, the necessity of expelling the resident inspectors, an act that in any event would require much preparation and consideration before it could be consummated.

Furthermore, it would seem that the generation of nuclear electricity ought to be entrusted to powerful industrial entities that would specialize in generation of nuclear energy. I would envision the nuclear enterprise operated by consortia—possibly like the Yankee Atomic Electric Company or TVA, possibly like the du Pont Company, which operated Hanford and all its supporting facilities during the war.

Finally, the entities that operate the enterprise will have to be invested with some degree of longevity, much as those that operate the dike system in Holland. A nuclear plant cannot simply be abandoned. Temporary wastes must be looked after and converted into permanent wastes. Such jobs require longevity, which undoubtedly means that the integrity of the institutions responsible for the nuclear enterprise will have to be underwritten by the government.

These attributes—colocation in relatively few energy centers operated by strong consortia, heavy security, professionalized cadre, and longevity of the responsible institutions—seem to me essential for the long-term acceptability of nuclear energy.

Solar Utopia vs Nuclear Utopia

The energy war is waged, as are many religious wars, over two differing conceptions of the future—the solar utopia and the electrical, i.e. nuclear utopia. Where the solar utopia is decentralized, the nuclear utopia is centralized. Where the solar utopia is nonelectrical, the nuclear utopia is electrical. Both utopias are conceivable. The solar utopia appears to me to be expensive and inconvenient; the nuclear utopia poses difficult social questions.

As seems to be the case in religious wars, no quarter is asked or given. Thus the attempts to devise an acceptable nuclear future, perhaps along the lines I have tried to outline, are met with hostility by the energy radicals. They prefer to view the problems of nuclear energy as intractable; but even if the problems could be resolved, the centralization of the nuclear utopia fills them with repugnance—they will have none of it.

It is ironic that Franklin Roosevelt in 1936 looked upon rural electrification, based on central power stations, as the key to a sort of Jeffersonian, decentralized society. Sheldon Novick describes Roosevelt's hopes for electrification:

> Decentralization, if and when it takes place, will break up the great conglomeration of people in sprawling . . . slums . . . and put them back on the land. Electric power, a vast unseen ocean of electric power that will run factory machines, light the countryside and bring relief from drudgery to the homes on the land, is for Roosevelt the seemingly certain instrument of this decentralization.[8]

And in his address to the World Power Conference in 1936 Roosevelt said:

> Sheer inertia has caused us to neglect formulating a public policy that would promote opportunity to take advantage of the flexibility of electricity; that would send it out wherever and whenever wanted at the lowest possible cost. We are continuing the forms of overcentralization

> of industry caused by the characteristics of the steam engine, long after we have had technically available a form of energy which should promote decentralization of industry.

Thus our perceptions of the evils of centralization have changed in the past; and they will change again in the future. It seems to me therefore that the exclusivity of the energy radicals—that is, their commitment to the Sun as the only solution—is short-sighted. Having been guilty in the past of predicting that nuclear electricity would cost ten times less than it now costs, I am profoundly skeptical of any estimates of the costs of an energy system to be deployed a decade or more in the future. Indeed, it may be that the solar utopia is so expensive that it could not penetrate the market—i.e., come into being—in competition with the nuclear utopia unless a political commitment were made foreclosing the nuclear option. To take such a step—that is, to destroy nuclear energy rather than trying in the most serious way to fix it, perhaps along the lines I suggest, seems to me to be utterly wrong. It seems particularly ill-considered when we still have so little grasp of the full costs—ecomomic, social, and political—of the solar alternative.

The most prudent future will undoubtedly use some combination of the two sources: solar energy where it makes thermodynamic and economic sense, and where its inherent intermittency can be dealt with; nuclear energy where it makes sense. The ultimate system will almost surely be a mixture—centralized and decentralized; electric and nonelectric; nuclear and nonnuclear. To wage a war now to dictate the shape of that unforeseeable future in the face of so many economic, technological, and social uncertainties can only be described as extremely dangerous. What we must do is press on with the research and development on both solar and nuclear energy that will resolve these uncertainties. What we need are peace treaties in the energy wars that will allow our energy system to develop along lines that make sense not only to proponents of a particular technology or of a particular social philosophy, but to the majority of the people whose descendants will inherit the energy system we bequeath them.

References

[1] A. B. Lovins, in *Soft Energy Paths: Toward a Durable Peace* (Ballinger, 1977).

[2] R. M. Rotty and E. R. Van Artsdalen, Thermodynamics and its value as an energy policy tool. Energy **3**, 111–117 (1978).

[3] W. van Gool. Limits to energy conservation in chemical processes. Institute for Energy Analysis, Oak Ridge Associated Universities.

[4] D. T. Spreng, 1976. Useful questions concerning energy accounting. Institute for Energy Analysis, Oak Ridge Associated Universities, ORAU/IEA (M) 76–77.

[5] C. S. Hollings, in *Future Strategies for Energy Development: A Question of Scale—1976, Oak Ridge, TN*, proceedings of a conference, ORAU-130.

[6] American Physical Society. Report to the APS by the study group on nuclear fuel cycles and waste management. Reviews of Modern Physics **50** (1978).

[7] N. C. Rasmussen, 1975. Reactor Safety Study: An Assessment of Accident Risks in U.S. Commercial Nuclear Power Plants. WASH-1400 (NUREG 75/014).

[8] S. Novick, in *The Electric War: The Fight Over Nuclear Power*. Sierra Club Books, p. 236.

A Family of Maxwell's Demons

Into 1980 fell the 109th birthday of Maxwell's Demon. The ingenious little fellow first appeared in Maxwell's *Theory of Heat*, published in 1871.[1] Maxwell wrote:

> Let us suppose that a vessel is divided into two portions, *A* and *B*, by a division in which there is a small hole, and that a being who can see the individual molecules opens and closes this hole, so as to allow only the swifter molecules to pass from *A* to *B*, and only the slower ones to pass from *B* to *A*. He will thus, without expenditure of work raise the temperature of *B* and lower that of *A*, in contradiction to the second law of thermodynamics.

In other words, a heat engine that produces work could be operated on this temperature difference perpetually, or at least until the demon got tired. That the resolution of this paradox might have something to do with information was hinted at by M. von Smoluchowski in *Lectures on the Kinetic Theory of Matter and Electricity*, published in 1914.[2] Smoluchowski wrote (p. 89): "There is no automatic, permanently effective perpetual motion machine, in spite of molecular fluctuations, but such a device might perhaps function regularly if it were operated by intelligent beings."

Based on the Keynote Address to the National Conference of the Association of Computing Machinery, Nashville, TN, 1980; published in Interdisciplinary Science Reviews 7 (1), 47–52 (1982). Reprinted by permission.

Szilard's Argument

Leo Szilard, one of the fathers of nuclear energy, first resolved the paradox by establishing a quantitative connection between information and entropy. In a remarkable paper first published in 1929 in the *Zeitschrift für Physik* and reprinted in 1964 in, of all places, the American journal *Behavioral Science* under the title "On the Decrease of Entropy in a Thermodynamic System by the Intervention of Intelligent Beings," Szilard showed that the second law of thermodynamics would not be violated if the entropy S of the system increased by an amount

$$\Delta S = k \ln 2 \tag{1}$$

(where k = Boltzmann's constant = 1.38×10^{-23} joules per degree Kelvin) every time the demon measured the speed of a molecule in order to decide whether or not to open the trap door.[3] This increase in entropy just balances the *decrease* in entropy associated with the appearance of a higher-speed molecule in the right-hand box. In this way, the paradox is resolved, since the entropy of the system informed-demon-plus-segregated-gas is the same as that of the system uninformed-demon-plus-unsegregated-gas. We can therefore quantify the information in the demon: per measurement, or per bit, it is

$$\Delta I = -k \ln 2 \tag{2}$$

the negative sign signifying that, for each increase in information, there is a corresponding decrease in the actual physical entropy of the gas.

Szilard's argument implies that the connection between information and entropy is intrinsic and real, not incidental and analogic. Pushed further, one would have to concede, as Eugene Wigner has explained to me, that entropy itself has anthropomorphic connotations. For example, as a red liquid and a blue liquid are mixed, their degree of mixing, and therefore the entropy of the system, depends on how well an observer can keep track of the smaller and smaller globules of each liquid. An observer with very sharp eyesight can keep track of individual globules that are too small to be seen by one with poor eyesight. For him, the two liquids are really not mixed, and, in principle, he could fish out the blue from the red. The entropy, as he measures it, would therefore differ from that of the observer with poorer eyesight.

Szilard's ideas appear in the work of C. Shannon[4] and later in that of L. Brillouin[5]; Shannon, entirely independently of Szilard, derived a formula equivalent to Equation (2). But Shannon, when he speaks of entropy of information, is using it in an analogic sense: he makes no direct connection with the classical, phenomenological concept of entropy,

$$\Delta S = \frac{\text{Amount of heat transferred in a reversible process}}{\text{Temperature at which heat is transferred}} = \frac{\Delta Q}{T}, \tag{3}$$

as a state variable characterizing a thermodynamic system. Indeed, according to M. Tribus,[6] Shannon attributes his use of the word entropy to John von Neumann, who pointed out to him that "no one knows what entropy really is so in a debate you will always have the advantage."

{Actually, as C. H. Bennett and R. Landauer showed in the 1980s [Scientific American **255**, 108–116 (1987); **253**, 48–58 (1985)], Szilard was slightly wrong in attributing to the act of measurement itself the entropy change given by equation (2). A closer analysis by these authors shows that it is the act of erasing prior information from the demon's memory, not the act of measurement itself, that entails the entropy change $-k \ln 2$. However, since the previous bit of information must be erased before a new bit can be acquired, we can still speak, somewhat loosely, of $-k \ln 2$ as the entropy change connected with each measurement (where "measurement" now includes both erasing prior information and imprinting new information).}

Duality of Entropy and Information

In this essay, I shall adopt Szilard's point of view—that on a molecular level there is an intrinsic connection, even a duality, between information and entropy: increase in information corresponds to a diminution of physical entropy. Moreover—if one does not count the energy required to acquire the information—intelligent intervention on a microscopic level allows one to extract more useful work than would have been possible without this intervention. But, as Brillouin explained in detail, energy must be expended in order to let the demon make his measurement. Thus, the entropy decrease in the demon requires an expenditure of energy:

$$\Delta Q = T\,\Delta S \approx 10^{-23} T \text{ joules per bit} \tag{4}$$

or, at room temperature, $\Delta Q \approx 3 \times 10^{-21}$ joules per bit. This expenditure of energy just balances the energy derived from the segregation of the molecules in the box. Thus, if we call E_A the extra energy made available by the demon's intervention and E_D the energy required for the demon to operate, then their ratio is

$$R = \frac{E_A}{E_D} = 1. \tag{5}$$

It will be my purpose to show that this almost bizarre connection between information and entropy has analogies also on a macroscopic, physical level and even on a social, institutional level. Moreover, as the level of aggregation increases, the energy saved by astute intervention of information processing devices—that is, by control—can greatly exceed the energy expended in operation of the devices, unlike Maxwell's Demon where, as seen in equation (5), the two energies are equal.

Why Entropy?

So far, I have written mostly about information or control and entropy. Yet a moment's reflection should make clear that it is entropy, not energy *per se*, that is of primary interest—that the energy crisis is a misnomer. The first law of thermodynamics states that energy can be neither created nor destroyed: in what sense, then, are we in an energy crisis, since the amount of energy always remains the same?

The answer, of course, is that we lack useful energy—that is, energy in a form that enables us to extract work, to heat a cold chicken, or to drive a chemical reaction. Various so-called state functions measure how much useful work can be derived from, say, the energy in a piece of coal, or a compressed gas, or a tank of hot water. These state functions include the Helmholtz free energy, F; the Gibbs free energy, G; and the engineer's availability, A. The appropriate state function depends upon whether the transformation is being performed at constant temperature, F; or at constant pressure and temperature, G; or in an environment at a fixed temperature and pressure that differs from that of the system performing the work, A. All these functions are linear functions of the entropy, and the coefficient of the entropy term is negative. Because the coefficient is negative, the higher the entropy of the system, the less useful work can be derived from it. Thus, insofar as man's transactions involve converting the energy in a chemical or nuclear or solar system into useful work, such transactions are most efficient—that is, can be accomplished with the least expenditure of useful energy—if the entropy of the system initially is as small as possible.

A simple example might make this clearer. Suppose an electric heater is energized by an amount ΔW_E of electric energy. The heater supplies an amount of heat $\Delta Q \approx \Delta W_E$ to a hot water tank, thus raising its absolute temperature to T. The electrical energy represented in the electric current flowing in the heater is by convention regarded as being work—that is, energy with zero entropy. If this electricity were run through a motor, all this electrical energy, except for frictional losses, could in principle be converted to mechanical work, ΔW_{M}. On the other hand, once the electrical energy is converted into an amount of heat ΔQ at temperature T, the entropy of the hot water has increased by an amount $\Delta S = \Delta Q/T$. A heat engine energized by the hot water could generate an amount of mechanical energy that is less than the energy in the electricity by at least the added entropy multiplied by the temperature T_0 at which the heat is rejected. Thus, the maximum possible work $\Delta W_{\max}$ supplied by the heat engine from the increment of heat ΔQ is

$$\Delta W_{\max} \leqslant \Delta Q - T_0 \, \Delta S = \Delta Q(1 - T_0/T). \tag{6}$$

We see that the higher the operating temperature T—that is, the lower the entropy change—the greater the amount of work we can extract from the increment of heat, ΔQ.

To take another example, suppose we are trying to heat a house 10 degrees above the outside temperature. From the point of view I have delineated, we ought to use as a heat source a relatively low-temperature source (say, a geothermal well) rather than a high-quality, low-entropy source (say a bucket of coal). Even though the amount of energy required to heat the house is the same in both cases, we are being wasteful in the second case because, in addition to heating the house, we could have extracted more useful work out of the low-entropy bucket of coal than out of the high-entropy geothermal well.

Macroscopic Maxwell's Demons

As stated, the crisis is not simply about energy: it is about energy of high quality—that is, energy at low entropy. The 17 million barrels of oil that flowed each day through the Straits of Hormuz before the Iran–Iraq war represented a critical supply of energy at low entropy. Our national interest in the United States, not to speak of the world's interest, required us to utilize this precious supply of energy at low entropy as efficiently as possible.

Until the advent of microcomputers and other ways of handling information, the path to higher efficiency, and therefore better utilization of our limited supply of oil, in conversion to mechanical work depended primarily on reducing the entropy change in the conversion by increasing the temperature of the working fluid—for example, the inlet temperature of the aircraft gas turbine has increased from about 760 K to 1600 K in the past 30 years. The theoretical Carnot efficiency of aircraft turbines has increased from about 30 percent to 60 percent, the actual efficiency from about 15 percent to about 30 percent.

Materials limit the improvements in efficiency that can be achieved through increasing temperatures. Fortunately, there is an entirely different approach to increasing efficiency, an approach in the spirit of Maxwell's Demon, that is not limited by the temperature that a turbine blade can withstand without failing. I refer, of course, to the use of microsensors and microcomputers to monitor and control the conversion of chemical energy into work in an engine, so as to minimize the increase in entropy in the process.

Recently, General Motors placed a full-page advertisement in *Parade* (6 October 1980), a Sunday newspaper magazine that circulates very widely. To quote:

> Computer Command Control . . . allows GM to achieve the highest Corporate Average Fuel Economy . . . . A solid-state electronic control module monitors oxygen in the exhaust, engine speed, and engine coolant temperature through three highly specialized sensors . . . . Analyzing this information at thousands of calculations per second, it then adjusts the air/fuel mixture in the carburetor to optimize combustion . . . .

I do not think I am stretching a point too much by referring to such microprocessor control systems as macroscopic Maxwell's Demons. According to E. F. Walker, head of the General Motors Electronics Department, the energy saved by the Computer Command Control is about twice the energy required to operate the system—that is, $R \approx 2$.

The General Motors Computer Command Control is one of an ever-growing class of information processing devices that are improving the efficiency of conversion of chemical energy to work. Although most improvements in automobile efficiency stem from reduction in weight of cars—the average miles per gallon for General Motors cars is expected to reach 31 by 1985—intervention of microprocessors, according to Daimler-Benz, might improve efficiencies in Mercedes engines by another 5 or 10 percent (personal communication, E. Schmidt).

Systems Maxwell's Demons

Another class of information-handling device that can save energy is exemplified by the automatic control systems for heating units in large buildings. These systems, as described in *Energy User News*, try to match use of energy with actual demand for energy. In this sense, they might be described as Systems Maxwell's Demons, as they make a whole system, consisting of user of energy and producer of energy, operate more efficiently. For example, with the Fabri-Tek system, a guest activates the heating or air conditioning in his room with his room key—in effect, only occupied rooms are conditioned. But such devices, unlike the GM Computer Command Control, impact on the guests' convenience, a point to which I shall return later. To quote from *Energy User News*, "Every time a customer would leave the room, he'd have to pull his key out, and when he came back the room would be cold... It takes 2 hours to warm a room from 60 °F to 70 °F. Not very many customers are willing to wait that long."[7]

Another example of how information control saves energy on a system scale is the Flight Management Computer: according to Robin Wilson of Trans World Airlines, use of such computers to optimize flight plans saves 2 percent of the total fuel used in the TWA system. Optimizing the flight plan for the Concorde saves even more fuel—5 percent—since the Concorde uses so much more fuel per kilometer than a subsonic plane.

The late Pat Haggerty of Texas Instruments, in his speech "Shadow and Substance," offered a far more expansive view of the possibilities for saving energy through intervention by microprocessors.[8] To begin with, he found a correlation between the use of Active Element Groups (AEG) and energy saved in automobiles and households. He defined an AEG as a transistor plus its associated circuitry, including diodes, resistors, capacitors, printed circuit boards, etc. From this correlation, he estimated that 53×10^{15} Btu of energy, out of an estimated total of 180×10^{15} Btu, could be saved in the United States in the year 2000 A.D., if 1.4×10^{13} AEGs could be installed by that time. In 1977, the U.S. semiconductor industry produced 4×10^{11} AEGs. Haggerty considered that, by 2000 A.D., this number would grow to 2×10^{16} AEGs; thus, only 1/1000 of the yearly production at that time would be used in a variety of ways for engine and process control and for a substitution of transfer of information instead of a transfer of people. I am unable to evaluate Haggerty's extraordinary prediction of how microprocessors will greatly reduce our energy demand. I suppose he himself would have conceded that the correlation on which he bases his speculations, automobiles and households, is probably insufficient to make such great extrapolations. Nevertheless, one cannot help but be impressed by such extraordinary optimism about the future of microprocessors.

Social Maxwell's Demons

I have described how information can be traded for energy in a microscopic Maxwell's Demon, although in that case the energy expended by the demon actually equals the energy saved; in a macroscopic physical Demon (GM Computer Command Control); and in macroscopic systems Demons (Fabri-Tek and Flight Management Computer). It is possible to push the hierarchy further: we can find many examples of how clever use of information can save energy on what might be described as a social scale.

Consider the effect of price on consumption of fuel. When oil cost $2 per barrel, industrial processes, as well as cars, were designed with little regard for their fuel efficiency. With oil costing $30 per barrel, the market has transmitted signals to innumerable users of gasoline that it is to their advantage to buy cars that are fuel-efficient. This in turn has sent signals to Detroit: design a fuel-efficient car. There are sophisticated buyers out there who will buy such cars.

We see this trade-off between knowledge and energy in several ways here: the buyer is more sophisticated about the importance of fuel efficiency, and so he demands an efficient car; the automobile manufacturer must exercise ingenuity in designing a more efficient car.

In a broader sense, the whole effort to conserve energy requires re-education of innumerable energy consumers. This re-education may come from exhortation or from the impetus of the marketplace.[9] The price mechanism apparently has worked: the United States used 8 percent less oil in 1980 than in 1979, and our oil imports dropped from 8.5×10^6 barrels per day in 1975 to 7×10^6 barrels per day in 1980. Though some of this has come about because of recession, or a mild winter, much of it must be attributed to the decontrol of oil prices.

It is evident that the ratio of energy saved to energy expended in putting the information system in place—in this instance, the decontrol of oil prices—must be enormous. Thus, I would put forth the conjecture that, if R is the ratio of energy saved by the intervention of an information-control system to the energy expended in the information system, then, as the scale of the intervention increases, so does the ratio R. I illustrate this as follows:

$R = 1$	for microscopic Maxwell's Demon,
$R \approx 2\text{–}100$	for microprocessor control,
$R \to \infty$	social interventions—for example, by market mechanisms.

The Role of Time

In the previous examples, I have suggested that information, primarily its manifestation as control, can be used to save energy; and that this way of increasing efficiency through use of, say, microprocessors is a more delicate and sophisticated approach than is the use of higher temperatures. Yet going to higher temperatures to increase efficiency can be, in a sense, subsumed under the general heading of information: we go to higher temperatures only by using higher-strength alloys, or by applying more powerful methods to the design of components. Development and deployment of such alloys is therefore itself a manifestation of the use of information. In a broader sense, then, both higher temperatures and the use of microprocessors are captured in the word *information*. Moreover, this trade-off between information and energy is reciprocal: clever use of information can save energy; or energy can be used to compensate for lack of information, as when a reactor engineer overdesigns a pressure vessel because he does not know just how to estimate the effect of radiation on its strength.

Daniel Spreng, a Swiss physicist, has pointed out that there is another great trade-off: this is between time and energy.[10] If we do things slowly, we use less energy than if we do them fast: haste makes waste. For this there is a good thermodynamic basis. Reversible processes are more efficient than are irreversible ones, but reversible processes go infinitely slowly. Indeed, all of classical thermodynamics assumes that we are immortal, since it treats only reversible processes. This trade-off between time and energy we see in many everyday ways: the American 55-miles-per-hour speed limit saves oil (energy) but it costs time; if we match perfectly the quality of energy at the end use (heating a room, for example) with the energy of the source, it would take all day to

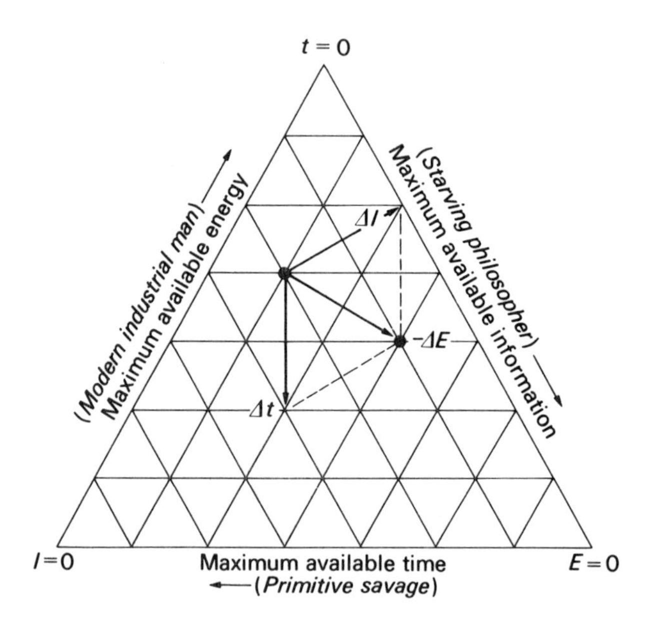

FIGURE 1. *The Spreng triangle.*

heat the room. Remember here the Fabri-Tek users' complaints that they have to wait for their room to be heated.

Spreng's Triad

Spreng thus conceives energy, time, and information as forming a triad: each, in some sense, can be traded off for the other two. He summarizes these trade-offs in what I call the Spreng triangle (Figure 1). To quote from Spreng,

> In this diagram each point inside the triangle represents a possible mix of energy, information, and time necessary to accomplish a certain task. When much available energy is employed, the point will be indicated at the upper left side of the triangle; if less energy is required, the point moves toward the opposite corner marked with $E = 0$. Two methods to accomplish a job with the same requirement of available energy will be represented by points lying on a line parallel to the side of maximum available energy use. Near the corners of the triangle are the situations of the starving philosopher (near $E = 0$) who employs much information and time to accomplish his task; of a primitive man, perhaps in a slash-and-burn society (near $I = 0$) who, because of little information,

requires much time and energy (a large area of forest) to meet his needs; and of the "industrial man" (near $t = 0$) who with much energy and information accomplishes things very quickly.

The diagram underlines the fact that energy conservation measures, ΔE, can be simultaneously the substitution of time, Δt, and the substitution of information, ΔI, for energy. For instance, increasing the efficiency by slowing down heat flows, i.e., by reducing the output rate or increasing the size of equipment, has to be done with expertise and know-how. Or, the educational process to disseminate energy conservation know-how requires time and patience.

As Figure 1 suggests, information and time can also be substituted for one another: if we have lots of time, we do not need information, but can apply trial-and-error approaches to solving a problem; if we have little time, we are happy if we know which is the fastest way. In fact, the main function of applying electronic devices in the last few decades has been to make more time available for saving labor and speeding up the flow of information.

The Centrality of Time

In the great energy debates, the role of time, unlike the role of information, seems largely to have been ignored. The extreme polarization of the debate was exemplified by the views of two presidential candidates, Barry Commoner and Ronald Reagan. According to Commoner, we can, even now, shift to a solar-based society and turn away from what are called hard technologies, the centralized electric power stations fueled with coal or nuclear sources. This shift implies far more vigorous conservation measures than we have already taken. Ronald Reagan, by contrast, saw few dangers in proceeding with our centralized energy production methods, including nuclear; and, although he accepted conservation and solar energy as desirable, he placed these at a lower priority.

On energy policy, I have much more sympathy with President Ronald Reagan's view than with Dr. Commoner's. I have two reasons for my position. First, the energy crisis we seek to resolve is not a general energy crisis; it is our dependence on unreliable foreign oil. From this viewpoint, replacing oil with nuclear- or coal-generated electricity makes sense. And even though the use of low-entropy electricity for heating houses—a job that can be accomplished with poorer quality energy sources—offends the sensibilities of energy radicals, it turns out to be not only convenient but economical at the present price of oil.

But the main reason why I favor a predominantly centralized, hard-energy system is that, on the whole, it is less intrusive and makes far fewer demands on our individual use of time than do Commoner's soft systems. The decentralized systems based largely on the Sun and wind are inherently less reliable than the centralized systems based on fossil or nuclear sources. A society based primarily on solar systems will tend to be an intermittent society: what we can do at any time will depend upon the vagaries of the weather. To be sure, farming and outdoor construction are already beset with such intermittency. But to impose such intermittency on all our activities seems to me to be an enormous price to pay for the alleged advantage of being able to control one's own private electrical system.

I am puzzled that the role of time, indeed its centrality, has been largely ignored in the energy debate. Yet both on a theoretical thermodynamic basis, as well as a practical basis, time is of the essence. One can even argue that time is our most valuable resource, and that the value of energy and, as Spreng says, of information, is to give us more freedom to allocate our time. To return to Spreng's triad, I have here speculated on how information saves energy: I have only touched upon how information saves time. Yet, in the long run, the impact of information on our freedom to allocate time may be even more important than its impact on our use of energy. In previewing the Computer Age, I would suggest that the reorganization of our use of time may be the most profound and lasting social effect of the extraordinary advances in the handling of information that have largely resulted from the work of ever more efficient computing machinery.

References

[1] J. C. Maxwell, in *Theory of Heat*, 3rd ed., republication of 1872 ed. (AMS Press, Inc., New York, 1972).

[2] M. von Smoluchowski, in *Lectures on the Kinetic Theory of Matter and Electricity* (1914).

[3] Leo Szilard, "On the decrease of entropy in a thermodynamic system by the intervention of intelligent beings, Behav. Sci. **9**, 4 (1964).

[4] C. Shannon, "A mathematical theory of communication," Bell Syst. Tech. J. **27**, 379, 623 (1948).

[5] L. Brillouin, in *Science and Information Theory* (Academic Press, New York, 1956).

[6]M. Tribus and E. C. McIrvine, Sci. Am. **225**, 179–188 (1971).

[7]"Lodging industry wary of automated energy controls," in *Energy User News* **5**, 39 (1980).

[8]P. E. Haggerty, "Shadow and Substance," presented at Semicon/West Annual Banquet, San Francisco, CA, 1977.

[9]The role of the market as an information-feedback system has been an important thread in economic theory; see, for example, T. C. Koopmans, "Efficient allocation of resources," Econometrica 445–465 (October 1951). I am grateful to W. Gilmer and D. Weinberg for pointing this out to me.

[10]D. T. Spreng, "On time, information, and energy conservation," ORAU/IEA-78-22(R) (Institute for Energy Analysis, Oak Ridge Associated Universities, Oak Ridge, TN, 1978).

Energy Policy and Mathematics

As befits a presentation before a mathematics society, I begin with a canonical representation of the "energy crisis." By this I mean a statement that captures the essence of the energy "problematique," and with which all other theorems and lemmas about energy must be consistent.

My universal statement consists of three parts:

1. Fossil fuel, represented by CH_x (where $x \cong 4$ for natural gas, $x \cong 2$ for petroleum, $x \cong 1$ for coal), comprises 66 percent worldwide—80 percent in the United States—of all mineral molecules used by human society.

2. Fossil fuel, extractable with a positive net energy balance, is about as abundant in the Earth's crust as the rare earth neodymium, and its distribution is uneven—for example, South America has very little coal, and the land mass of Europe has very little oil.

3. Extractable CH_x, though rare by global standards, is still about 15 times more abundant than the carbon found in the atmosphere as CO_2 (10,000 $\times 10^9$ tons of C as CH_x in the Earth's coal, 700 $\times 10^9$ tons of C as CO_2 atmosphere).

From the first statement follows the proposition that the extraction of CH_x will probably continue to be a large-scale activity; but, accord-

---

Presented as the Keynote Address, Society for Industrial and Applied Mathematics 1978 Fall Meeting and Symposium on Sparse Matrix Computations, Knoxville,TN, October 30, 1978; published in SIAM Review **22** (2), pp. 204–212 (1980). Reprinted by permission.

ing to the second statement, we must also dole CH_x out sparingly: in short, we must conserve. In any event, demand in the long run will outstrip supply, and the price of CH_x will rise inexorably. From the third statement we conclude that there may be an absolute limit to the amount of fossil fuel that can be burned without jeopardizing the Earth's climate, and this limit may be much smaller than the total extractable fuel. Many, if not most, of those who have studied the problem now believe that a doubling of the CO_2 in the atmosphere—which might follow from the burning of 10 percent of the globe's CH_x—could raise the average temperature of the Earth's surface by about 2 °C. (See, for example, Reference 6.)

The CO_2 catastrophe, if indeed it exists, is the ultimate in environmental constraints on the use of fossil energy. There are, of course, other environmental constraints—SO_2, radioactivity, local heating—on the use of energy. In addition to the environmental constraints there are other constraints, both on the use of energy and on its conservation. For example, if energy-conservation measures are implemented too rapidly, then because there is at least a partial trade-off between energy use and economic growth, economic growth is likely to diminish. And energy can be traded off for time, a point first brought out by D. Spreng of Swiss Aluminium, Ltd., in Zurich.[1]

Energy policy seeks to strike a politically acceptable balance between the use of energy and the constraints on its use imposed by the impact of energy on economic well-being, on the environment (both on a micro scale and on a global scale) and on our freedom in the use of time.

Energy and Time

Of these complex trade-offs, I regard energy and time as the most fundamental. This arises from underlying thermodynamic principles: generally, the faster a process goes, the more entropy it generates, and the more irreversible it is. Maximum efficiency requires thermodynamic equilibrium at each stage; and thermodynamic equilibrium exists only in processes that go infinitely slowly.

In practice, the situation is more complicated. For example, an automobile runs more efficiently at 35 miles per hour than it does at 10 miles per hour because some energy is used to idle the engine. But, beyond a certain speed, the efficiency falls off rapidly: haste indeed makes waste. The 55-miles-per-hour speed limit conserves energy since autos are more efficient at 55 miles per hour than at 75 miles per hour;

on the other hand, the society pays for this gain in energy (actually, free energy) by losing time.

Of all things precious, time, I suppose, would be ranked at the very top by most of us. As mortals, we know that our time on Earth is limited, and we are disciplined to use our time, not waste it. When we must choose between the two scarce commodities, time and energy, in our Western culture we are inclined to save time by using energy: we use personal cars, not mass transit; we use power boats, not row boats; we use electric dryers, not clotheslines. We are in a hurry. We pay for this with energy, not to mention ulcers.

This trade-off between time and energy is manifest in the utilization of intermittent energy sources, such as the Sun. Intermittency, after all, usually means a loss of time: An unexpected interruption of power in an aluminum plant will cause the plant to shut down. To overcome the intermittency inherent in an energy source, such as the Sun or wind, one must provide storage. But this requires additional energy, both to fill the storage (and to empty it, since the storage is never 100 percent efficient) and to build the storage system. To overcome stochastic intermittency and thus to regain a measure of control over the use of time, we must pay a price in additional energy.

I realize that there are other factors—capital, labor, information, materials—that can be traded for energy. I stress the importance of time, not only because I view it as most fundamental but because its significance in this context has been largely ignored.

Energy and Economics

It is customary to reduce most of the arguments over energy policy to questions of economics. The basic question is the relation between energy and economic growth. Just a few years ago, many energy experts held that the ratio of energy use to gross national product (E/GNP) was a constant of nature, at least in a given country. But closer examination of the data suggests that, in the United States, the ratio E/GNP has generally been falling (Figure 1).

The aim of much energy analysis is to estimate the future demand and supply for energy and to relate this to the other measures of economic well-being, particularly the GNP. The early projections of energy, for example those of P. Putnam in 1953,[2] were little more than extrapolations of historical trends; generally, such extrapolations have led to energy demands that were much too high. The most successful of the early projections of energy demand was that of S. Schurr *et al.*[3]

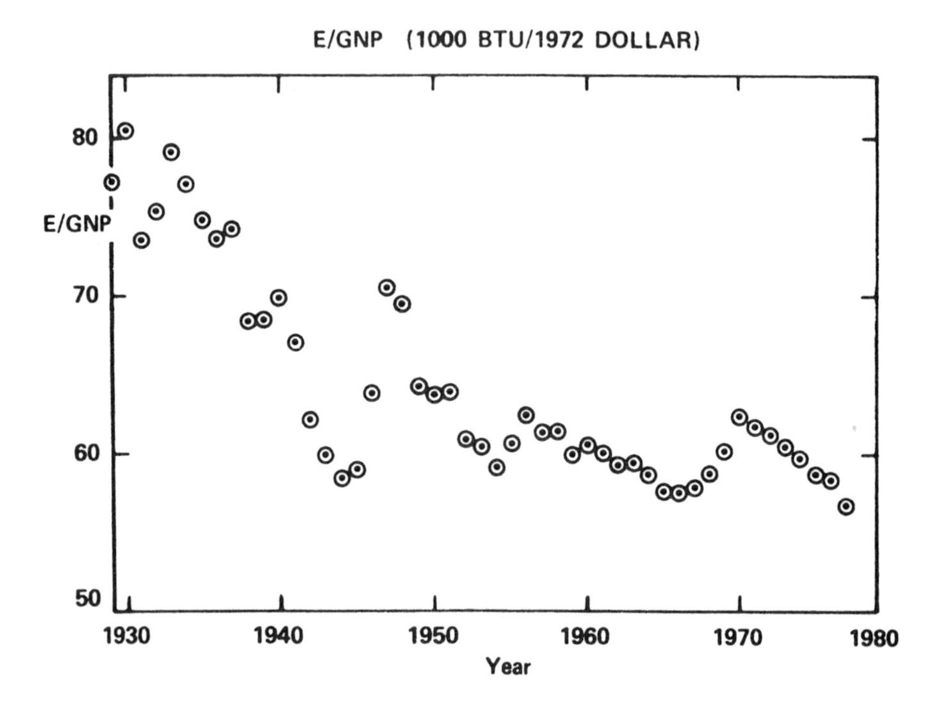

FIGURE 1. *E/GNP ratio for the United States.*

published in 1960. Using a judicious combination of extrapolation from historical trends and common sense with respect to saturation of energy demands, these authors came out with an aggregate projection for 1975 of 75 quads (1 quad $= 10^{15}$ Btu)—a remarkably accurate forecast, since the actual energy demand in 1975 was about 71 quads, and in 1976 it was 74 quads.

In the Schurr book, there were almost no equations. By contrast, it has become increasingly common in the recent literature of energy futurology to use elaborate economic models. These can range from extremely detailed models that seek to encompass the entire economy and relate energy to the growth in each economic sector, to much simpler global models—for example, the two-sector model of Edmonds and Reister.[4] In the latter model, the output of the economy is dependent upon the value of four variables: labor (L), capital (K), energy (E), and materials (M). The energy sector and the materials sector are considered separately from labor and capital, the latter being viewed as stocks, the former as flows. The main purpose of the model is to provide forecasts of energy demand and GNP; in particular, to estimate the economic impact of the transition from inexpensive energy sources to more expensive energy sources.

In general, the economic models seem to agree: the economy can withstand substantial increase in price of energy without serious impact, but only if the price increase is slow. Yet, one must be cautious: as Reister and Edmonds say, ". . . we cannot know the future. Consequently, we cannot know either the exogenous variables" (the parameters upon which the model is based) "or the endogenous variables" (such as GNP and energy demand generated by the model).

Though the economic models have a certain elegance, they depend very strongly on the price elasticity of demand for the most important energy carriers. Thus, the Project Independence model of 1974 derived an elasticity of demand for liquid fuels in transport that was rather high—about 0.75 (that is, a 1 percent increase in price of liquid fuel would reduce demand 0.75 percent). This can have far-reaching political implications: it means that, to reduce import of foreign oil (a constant aim of energy policy), all one need do is raise the price of gasoline—that is, remove controls on the price of crude oil. This policy is in sharp contrast to an alternative: namely, to reduce use of oil by rationing and allocation. One has here in confrontation two separate approaches to reducing import of foreign oil: the first is to free the marketplace of control, the second is to impose stiffer control (rationing).

The Ford administration, which was in charge when the Project Independence model was first developed, opted for control by freeing the market; Congress and, to a lesser degree, the Carter administration have been less willing to depend so heavily on the marketplace—Congress because the marketplace, although efficient, does not take care of poor people who would be hard hit by higher gasoline prices; the Carter administration perhaps because the elasticity coefficients derived in the 1976 Federal Energy Administration model have been lowered from 0.75 to around 0.5, and the lower the elasticity, the less effect an increase in price is likely to have on aggregate demand. [One year after I presented this paper, the Carter administration moved much more strongly toward decontrol.]

I mention this point to illustrate the power of mathematical modeling in affecting government policy. Indeed, the use of the Project Independence mathematical model to determine a political position (mediation by the market versus political intervention) represents, as far as I know, an unprecedented dependence of political policy on the output of two CDC 7600 main-frames!

Energy and the Environment

Mathematical modeling is used very extensively in predicting the impact of various energy sources on the environment. For example, the distribution of carbon dioxide in the atmosphere is a matter of considerable moment, and mathematical or systems ecology has become a technique for estimating this distribution. The general idea of applying systems ecology to estimate the distribution of pollutants is simple: one sets up equations linking the concentration of the pollutant in different interacting biota and different environmental "compartments." Among different species and environments there is assumed to be an exchange that is characterized by various exchange coefficients. For example, to go back to the concentration of CO_2 in the atmosphere, this concentration is governed by a set of coupled first-order differential equations that link it to the concentration of carbon in the oceans, in the terrestrial biota, and in the sediments, and to the amount of CO_2 injected into the atmosphere by human activities. In one such model by Olson, Pfuderer, and Chan,[5] there are, in all, 18 simultaneous equations. One of the key parameters is the response of the land biota to the atmospheric CO_2; if this is very large, the CO_2 concentration would be buffered by increased growth of plants; if small, the CO_2 would not be so buffered. Most authors agree that this exchange is too small to prevent CO_2 from building up, if we continue to burn fossil fuel at an ever increasing rate.

Predicting the concentration of CO_2 in the atmosphere as the result of our injecting the products of combustion is only the first step. One must also estimate what the effect of CO_2 on the state of the climate will be. In first order, the effect is simple. The Sun radiates to the Earth as a black body at a temperature of 6000 K; the atmosphere is largely transparent to this radiation. The energy is absorbed by the Earth and reradiated at a temperature of about 255 K. The 15 μm absorption band of CO_2 falls near the peak of the re-emitted radiation spectrum—in effect, the emissivity of the Earth is reduced by the presence of the CO_2. This decrease in emissivity is, in zeroth order, balanced by an increase in the Earth's radiative temperature: this is the CO_2 greenhouse effect.

This is only a schematic first approximation. In higher approximations, one must take account, for example, of the reduction in Earth's albedo caused by melting the arctic ice: this tends to increase the absorption of incident solar radiation, and thus to enhance the greenhouse effect, as Budyko[6] pointed out in 1969. But this is still a gross oversimplification: the actual atmosphere is three-dimensional; there is interaction with the oceans; there is feedback caused by the increased cloud cover, and consequent change in albedo, as well as many other intricate feedbacks. Estimating the climatic effect of CO_2 is now done by means of very complex general circulation models. Such models simulate the hydrodynamic behavior of the atmosphere–ocean system, with as much accuracy as the size of the computer will allow. Most of the complex global climate models and the simplest Budyko-type zero-dimensional models give about the same result: that a doubling of CO_2 concentration will raise the average surface temperature by about 2 °C—but the uncertainties in these estimates are very large.

The Role of Mathematics in Energy Policy

I have indicated how mathematics is involved in our increasingly sophisticated attempts to predict the future in two instances: the future of energy growth and the environmental implications of energy growth. I could give other examples of how mathematics enters into energy policy, for example, in helping us decide on the epidemiologic consequences of low-level insults in the atmosphere, or the estimation, from theories of mineralization, of how much uranium we are likely to find.

Now it seems to me that questions of this sort, often, if not always, have an inherently undecidable quality to them. It may be that the undecidability resides in the fact that one is dealing with extremely rare events, such as the biological consequences of low-level environmental insult, and frequency of occurrences of such rare events can never be tested empirically. Or it may be that the necessary data, such as elasticity coefficients, are not at hand, or that the underlying issues simply cannot be encompassed in the mathematics. (Could one predict the taxpayer's revolt or the toppling of the Shah?) Or, as described in the excellent article by H. L. Swinney and J. P. Gollub,[7] the undecidability may be intrinsic in the mathematics. Thus, E. N. Lorenz[8] first pointed out that the behavior of a two-dimensional fluid under so-called Rayleigh–Bénard flow can be described, in lowest approximation, by the set of equations

$$dx/dt=\sigma\,(y-x),$$

$$dy/dt=rx-y-xz,$$

$$dz/dt=bz+xy,$$

"where $x(t)$ is proportional to the amplitude of the convective motion, $y(t)$ is proportional to the temperature difference between ascending and descending currents and $z(t)$ is proportional to the deviation of the temperature profile from linearity." Let me now quote verbatim from Swinney and Gollub:

> The equations of the Lorenz model represent an extremely severe truncation: An infinite set of coupled ordinary differential equations has been reduced to three equations. The model therefore cannot be a realistic one for Rayleigh–Bénard convection significantly beyond $r = 1$. However, the Lorenz model is of intrinsic interest because of its fascinating mathematical properties. With the values of σ and b chosen by Lorenz, the system of equations has a stable solution $x = y = z = 0$ for $r < 1$, and two stable solutions,
>
> $$x=y=\pm[b(r-1)]^{1/2}, \quad z=r-1$$
>
> for $1 < r < 24.1$. In the "phase space" spanned by the amplitudes x, y and z (not to be confused with the spatial coordinates), the state of the system at any time is given by the point $[x(t), y(t), z(t)]$. For $r < 1$, all phase trajectories (representing different initial conditions) asymptotically approach a single point, the origin. Similarly, if $1 < r < 13.9$, all trajectories asymptotically approach one of the two stable solutions (which correspond to left- and right-handed rolls).
>
> Between 13.9 and 24.1, James Kaplan and his associates have found a complex transitional behavior; there is a metastable state, which can appear chaotic for arbitrarily long times, but the system finally settles down to a steady state. For $r > 24.1$, all trajectories are attracted toward a subspace on which they wander erratically *ad infinitum*. This is an example of a strange attractor. Many scientists are now investigating the properties of strange attractors for different nonlinear systems.
>
> The Lorenz model demonstrates that chaotic dynamics can be inherent in deterministic equations; thus the model supports the hypothesis that the complex dynamical behavior called turbulence is inherent in the hydrodynamic equations rather than being caused by random influences or a breakdown of the equations. Moreover, the Lorenz model shows that in a strongly nonlinear system *only three* degrees of freedom are needed to produce chaos! It has been conjectured that chaos in a system with many degrees of freedom (such as a fluid flow at large Reynolds numbers) is not qualitatively different from chaos in a system with a few degrees of freedom. Investigations of nonlinear model systems with a few degrees of freedom are now under way in many areas of science.

The hope is that such studies will lead to general insights into the chaotic dynamics of nonlinear systems.

This quality of undecidability—that is, the possibility that the evolution of the behavior of a nonlinear system depends discontinuously on the initial conditions, so that an infinitesimal change in an initial condition can lead to entirely different long-term regimes—I suppose bodes poorly for the role of present-day mathematics in policy analysis. And this possibility is catching: As Swinney and Gollub point out, such undecidability could be inherent in ecology, geophysics (including climatology), population dynamics, and economics. I call to your attention a meeting of the International Institute for Applied Systems Analysis (IIASA) about two years ago, at which econometricians, ecologists, climatologists, energy analysts, and topologists exchanged views as to their intersecting interests, and the existence or nonexistence of "resilient" systems.[9]

At least some economists share this pessimistic view. Thus, H. Y. Wan, Jr., in a paper titled "Causally Indeterminate Models via Multivalued Equations," states, "Economics is for prediction. But even with full information and in the absence of randomness, can one *always* predict a *unique* outcome? My answer is no. . . ."[10]

I am not a sufficient mathematician to understand the implications of all this. But, for the moment, let me play the devil's advocate and insist that there probably are undecidable propositions in energy analysis; indeed, that much of energy analysis is beset with propositions, such as proliferation, or the amount of uranium that can be found in the next 50 years at a price below $100 per pound, that are intrinsically or practically undecidable. Where does this leave energy policy? Is energy policy a hopeless undertaking?

I do not think so, since I believe we can do a great deal toward charting future courses that are based on propositions that are decidable. Thus, consider two propositions about the breeder reactor:

1. We do not know how much uranium there is nor what future energy demand will be, and therefore we do not know when we shall need the breeder.

2. We do not know what the breeding ratio or the cost of a developed breeder will be.

The first proposition, I assert, is inherently undecidable, since it requires not only a complete uranium mapping of the Earth but also an estimate of what the demand for nuclear power will be in the year 2000. By contrast, the uncertainties in the second proposition are removable—we need simply to design and build several breeders, and then we shall know the answer.

Mathematicians are, of course, called upon to help decide propositions of both types: but in the first type, their help may be to little avail; in the second type, it may be crucial. Indeed, most of the papers presented to this Society on the applications of mathematics in energy have to do with situations of the second type: development of technologies and models that admit of deterministic, nonchaotic solutions.

Can Energy Policy Be Based on Decidable Propositions?

To what extent can we base an energy policy on decidable propositions? What I have in mind is the possibility of getting around undecidable uncertainties at least to a degree: by developing and deploying technologies that circumvent those uncertainties—that is, by means of technological fixes that involve a minimum of uncertainty—or, alternatively, by recasting the whole policy issue so that it no longer depends upon an undecidable proposition.

Let me give a couple of examples. I have several times referred to the CO_2 catastrophe: but I have left the impression that its reality may itself be an undeterminable proposition. Is there a way of finessing the issue through technology? I see two possibilities: the most obvious is to sequester CO_2, perhaps as C. Marchetti suggests,[11] by removing N_2 from air *before* combustion (in which case the volume of gas to be handled is reduced fivefold). The second is to get on, far more vigorously than we are now doing, with the nonfossil alternatives that do not add CO_2 to the atmosphere—particularly fission, but also fusion and solar. There *is* urgency to get on with an answer to the decidable proposition—how expensive energy from breeders is—because we may want to deploy them very rapidly. Thus, we can finesse inherent uncertainties in energy policy by providing technological options; and the more technological options we have available, the less, in principle, will we have to depend on the answers to undecidable propositions.

More broadly, technology *per se* can circumvent uncertainties inherent in social engineering. Many of the uncertainties in our energy-economic models, as well as in our policies themselves, reside in the unpredictability of individual human behavior, a point brought out by Wan in his paper.[10] Now it is commonplace to suggest that the main element of a national energy policy is conservation. Yet conservation, by its very nature, usually depends on a myriad of individual decisions—people deciding to ride in car pools, or to turn their thermostats down, or to observe the 55-miles-per-hour speed limit. By contrast, increase in supply, say through synthetics, requires far fewer individual decisions. Once the government decides to spend 125 billion dollars for synthetics, there is little reason to doubt that large amounts of synthetic oil will be produced. Thus, increased supply, through technology, affords a more certain path toward reducing import of oil than does conservation. This basic point seems now to have been sensed both by Congress and by the President in their recent decisions to stress supply. It is not that conservation is undesirable; it is, rather, that conservation is more uncertain than increase in supply, if the latter is based on demonstrated technologies.

[This view, which I expressed in 1978, seems hopelessly out-of-date in 1989. Most of the "demonstrated" supply technologies—oil, coal, and nuclear—are under attack by a public worried about radioactivity, oil spills, and acid rain. And, indeed, the Carter Administration's attempt to introduce synfuels failed, partly on economic grounds, partly on environmental grounds. Were I writing this paragraph today, I would be much more even-handed in respect to conservation and supply technologies. As I point out in the next essay, *Energy in Retrospect*, conservation, largely mediated by the marketplace, has been a great success, whereas technologies that in 1979 I thought were "demonstrated"—particularly nuclear power—have fallen into public disfavor. I cannot believe these technologies will be forever proscribed by a skeptical public; rather, they pose a challenge to the technologists to make them, once more, fully acceptable to the public.]

It would be quite unbecoming for the director of an energy think tank to leave the impression that energy policy analysis is inherently useless since its most important questions are probably undecidable. This is surely a large overstatement: there are decidable questions in energy policy, for example, the feasibility of the breeder, or the general proposition that higher prices dampen demand. In any event, the systematic examination of alternatives certainly gives analysts valuable insights as well as places plausible limits on the consequences of actions, and these can be conveyed to policy-makers. But I believe we must maintain a kind of humility. I have seen how elaborate predictions—for example, with respect to the cost of nuclear energy—have turned out to be wrong by factors of 5 or more. No matter the elaborateness of the mathematical model, where the underlying mechanisms are not susceptible to mathematics—as I suspect is the case in econometric forecasting, especially when the forecasts are expected to hold for 20 or 50 years—there will be uncertainties. In a basic way, we cannot predict the uncertainties our children will encounter. What we can do is provide them with technological options that can be used to deal with the future energy situation that we in 1978 can only speculate about, but that they will face in its full reality.

References

[1] D. T. Spreng, "On time, information, and energy conservation," ORAU/IEA-78-22(R) (Institute for Energy Analysis, Oak Ridge Associated Universities, Oak Ridge, TN, 1978).

[2] P. C. Putnam, in *Energy in the Future* (Van Nostrand Company, New York, 1953).

[3] S. Schurr *et al.*, in *Energy in the American Economy, 1859–1975* (Johns Hopkins Press, Baltimore, MD, 1960).

[4] D. B. Reister and J. A. Edmonds, "A general equilibrium two-sector energy demand mode," in *Modeling Energy-Economy Interactions: Five Approaches*, edited by Charles J. Hitch, Research Paper R-5 (Resources for the Future, Washington, DC, 1977), pp. 199–246.

[5] J. S. Olson, H. A. Pfuderer, and Y.-H. Chen, "Changes in the global carbon cycle and the biosphere," ORNL/EIS **109**, Oak Ridge National Laboratory, Oak Ridge, TN, Sept. 1978.

[6] M. I. Budyko, "The effect of solar radiation variations on the climate of the earth," Tellus **21**, pp. 611–619 (1969).

[7] H. L. Swinney and J. P. Gollub, "The transition to turbulence," Physics Today, 41–49 (Aug. 1978).

[8] E. N. Lorenz, "Climate determinism," Meteorological Monographs **8** (30), 1–3 (1968).

[9] "Analysis and Computation of Equilibria and Regions of Stability–With Applications in Chemistry, Climatology, Ecology, and Economics, Record of a Workshop," CP-75-8, edited by H. R. Grümm (International Institute for Applied Systems Analysis, Laxenburg, Austria, 1975).

[10] H. Y. Wan, Jr., "Causally indeterminant models via multi-valued equations," in Bifurcation Theory and Applications in Science Disciplines, Ann. New York Acad. Sci. **316**, 530–544 (1979).

[11] C. Marchetti, "On geoengineering and the CO_2 problem," RM-76-17, International Institute for Applied Systems Analysis, Laxenburg, Austria, March 1976.

Energy in Retrospect

Before fission was discovered, energy policy was not a central issue in America. After all, ours was a country blessed with enormous reserves of fossil fuel, and we could hardly conceive of a day when the United States would be importing 30 percent of its oil. To be sure, Drake's well came just as whale oil was becoming scarce; and, prior to the discovery of the East Texas field, such alternatives as shale oil were pursued. But, by and large, the problem of energy in its broadest aspect had not become part of the Federal government's agenda.

The discovery of fission, which was widely regarded as the ultimate answer to the problem of energy, focused the government's attention on energy. It was as if, with the solution in hand, we became aware of the problem. Thus, in a 1953 report sponsored by the Atomic Energy Commission,[1] Palmer Putnam argued that a *prudent* custodian of the world's energy future should assume that energy demand would grow exponentially and that energy supply would turn out to be lower than the expansive estimates of supply then current. Though Putnam's maximum plausible world population by 2050 was only 6×10^9 people, his growth rate of energy use per capita of about 3 percent/year was very high; this led to his "maximum plausible" annual demand of energy of 436 quads by 2000 A.D., of 2650 quads by 2050 A.D. (1 quad $= 10^{15}$ Btu)! No wonder Putnam concluded that the world must get on with the development of all energy sources, especially nuclear power and solar energy (which he regarded as too expensive), as well as improving efficiency of utilization of energy. Incidentally, Putnam was the

Presented at National Academy of Engineering Symposium, Beckman Center, Irvine, CA, 1988.

first of the energy futurologists to call attention to the implication of the greenhouse effect for energy policy.

Putnam's report echoed, in apocalyptic tone, the Paley Commission Report[2] of 1952. This too warned that serious shortages in energy supplies could develop; but, by and large, Paley went unheeded. The 1950s and 1960s were periods of energy euphoria, although a few voices, notably M. King Hubbert, warned that the United States would become an oil importer by the 1970s.[3]

The euphoria reached its zenith with AEC's 1962 Report to the President on Civilian Nuclear Power, projecting some 734 gigawatts of electricity from nuclear power by the year 2000*; and the 1964 interagency study of energy research and development, which placed fission into a broader context of energy sources. This report found " . . . no ground for serious concern that the Nation is using up any of its stocks of fossil fuel too rapidly; rather there is the suspicion that we are using them up too slowly . . . we are concerned for the day when the value of untapped fossil-fuel resources might have tumbled . . . and the nation will regret that it did not make greater use of these stocks when they were still precious."[4] Despite this rosy estimate of our energy future, the interagency committee urged that the government expand research on long-range energy sources, both nuclear and nonnuclear.

These studies belong to what I would call the "pre-Cambrian" period of energy policy. During this period, an overall energy policy hardly seemed very relevant; and as for government-sponsored energy research and development, this was almost entirely preempted by the all-powerful Atomic Energy Commission and the Joint Committee on Atomic Energy. Although an Office of Coal Research had been set up in 1971, nuclear energy strongly dominated the government's thinking about the future of energy.

President Nixon's price freeze in 1971, followed by the Arab oil embargo in 1973, marked the beginning of the modern era of energy policy. The United States was then importing 6 million barrels per day (mmbd) of oil, and independence soon became the aim of U.S. energy policy. Thus, Dixy Lee Ray, chairman of the AEC, reported to the President in 1973 that we could achieve energy independence by 1985—but only if we conserved the equivalent of 14 quads (7 mmbd oil equivalent) out of a total annual demand of 100 quads[5]; and Project Independence in 1974 claimed we could achieve energy self-sufficiency

*This represented about 30 percent of a total projected energy demand of 135 quads.

by 1985 at an annual energy consumption of 96.3 quads, if oil prices rose by 20 percent and we conserved about 8 quads.[6]

These estimates of future demand were on the low side. Most forecasters at the time were predicting annual energy demand by 1985 of around 115 quads. Even Amory Lovins,[7] that archexponent of limited growth, was predicting 90 quads—a number close to the Ford Foundation's "Zero Growth" scenario.[8] Only the National Academy's Committee on Nuclear and Alternative Energy Sources[9] (CONAES), in its heavy conservation scenarios, spoke about demand for energy remaining constant, or even falling—but CONAES characterized its extreme conservation scenario as "very aggressive, deliberately arrived at reducing demand requiring some lifestyle changes." I don't think CONAES took this scenario very seriously.

In those days, several energy analysts took, as a rule of thumb, that the number of quads equalled the last two digits of the calendar year—78 quads in 1978, 79 in 1979, and so on. But the reality turned out very differently. Who, in 1973, would have predicted that the total amount of energy used in 1986 would be only 74 quads, the same as in 1973? Let energy forecasters practice their precarious art with humility!

The Energy/GNP Ratio

In the early 1970s, many of us were convinced that the ratio of energy to Gross National Product, GNP—or to Gross Domestic Product, GDP—was a constant, as indeed it was from 1945 to 1975. We seem to have forgotten that the E/GNP ratio had been falling from 1920 to 1940. The constancy of this ratio from 1940 to 1970 concealed the secular trend toward higher efficiency. This improvement in energy efficiency was evident in the entire Organization for Economic Cooperation and Development (OECD) (Table 1): between 1966 and 1970, the elasticity of demand of energy to gross domestic product, $\epsilon \approx \Delta \ln E / \Delta \ln GDP \approx 1.4$ (that is, a 1 percent rise in GDP was accompanied by a 1.4 percent rise in energy demand); between 1980 and 1984, $\epsilon \approx -0.2$ (a 1 percent rise in GDP was accompanied by a 0.2 percent *fall* in energy demand). Although the energy demand in the less developed and newly industrialized countries continued to expand, the entire non-Communist world has become considerably more energy efficient: $\epsilon \approx 1.3$ in 1966–1970; $\epsilon \approx -0.5$ in 1980–1983.

The other development has been the continued electrification of the United States and of the world. In 1968, some 18 percent of primary energy in the United States was converted to electricity; by 1987, this

TABLE 1. Growth rates of the demand for energy and the GDP of the non-Communist world.

		1966–1970	1970–1975	1975–1980	1980–1985
OECD	Energy (%)	5.8	1.3	1.7	– 0.2
	GDP (%)	4.2	3.0	3.5	2.1(– 1984)
	Elasticity	1.4	0.4	0.5	– 0.2(– 1984)
Non-Communist world	Energy (%)	6.0	1.9	2.6	0.7
	GDP (%)	4.5	3.6	3.9	1.4(– 1983)
	Elasticity	1.3	0.5	0.7	– 0.5(– 1983)

Source: BP Statistics: IMF-IFS, as quoted in MITI, p. 59, *Ministry of International Trade & Industry, Japan* (no date).[10]

fraction had doubled. The figures for the non-Communist world are similar. Moreover, the elasticity ratio of electricity to GNP does seem to be fairly constant, at least for the past 40 years.

Thus, the great realities of energy in the post-embargo world have been, first, the extraordinary flattening in the demand for energy in the developed world, which implies an unexpected decoupling of energy and GNP, and, second, the continuing electrification of the world and the remarkable correlation between electricity and GNP.

How can we explain this extraordinary diminution in the growth of energy demand—a diminution so extreme as to call into question the usefulness of energy-demand forecasting? Four schools of thought have arisen to explain away this discrepancy.

First are the *energy economists*. For them, the reduction in energy demand simply reflects both the lowered rate of economic growth and the increase in the price of energy. Even today, the average price of all energy is some two to three times the price of energy 15 years ago—little wonder that demand has abated. Moreover, since all this has resulted from the operation of the market, the energy economists, by and large, support nonintervention as our basic energy policy. The market has worked; let it continue to work.

A second group of analysts are the *structuralists*. For them, an important, if not dominant, reason for energy uncoupling from GNP is a change in the structure of our economic activity—the shift from manufacturing, mining, and agriculture to service, coupled with a saturation in some end uses (for example, television sets and refrigerators). Since service, by and large, is less energy intensive than manufacturing, energy demand has flattened. Thus the structuralists' policy implication would be: encourage further shift to service—energy will then take care of itself.

Third are the *conservationists*. These include the doctrinal conservationists, who regard conservation of energy to be a transcendent human purpose; and the technical conservationists, who simply insist that the technology of efficiency, both in end use and in energy production, has improved greatly and can be further improved. Moreover, these improvements often result in lower costs overall. For them, much of the reduced ratio of E/GNP reflects the adoption of more efficient technologies—in part prompted by the rise in energy prices, in part prompted by the widespread acceptance of a conservation ethic. Energy policy must therefore stress increased technical efficiency; but, at least for doctrinal conservationists, the government must mandate efficiency standards, such as Combined Average Fleet Efficiency (CAFE) and Building Efficiency Performance Standards (BEPS), as well as promote broad acceptance of the principle that conservation *per se* is ethically superior to any alternative.

Fourth (and finally) are the "*electro-niks*." For them, the increase in electrification and the reduction in E/GNP are not coincidental. Rather, electrification of industry is *per se* a powerful catalyst of increased productivity. As industry electrifies, whether or not the electrification is itself energy-efficient, industry becomes more productive. Electrification increases the denominator of the E/GNP ratio rather than diminishing the numerator, but the result is a decoupling of energy growth and economic growth. Energy policy for the electroniks is: encourage electrification, since an electrified society is an energy-efficient society.

There is truth in the views of all four groups of analysts. Nevertheless, although analysis of energy supply and demand is very much more sophisticated now than it was at the time of Project Independence, all of us must retain a basic skepticism about our ability to predict, much less mold, our energy futures even to the year 2000. We are confronted with an intrinsic dilemma: energy policy, insofar as it requires decisions today that affect the world 10, 20, even 50 years hence, must rely on our visualization of that future world; yet, as the inglorious history of our past energy projections has demonstrated, we cannot know that world. The main lesson from our experience is that, insofar as possible, we must try to formulate energy policies that finesse these uncertainties, that are resilient to surprises. Is this a realistic possibility?

Increasing Supply and Reducing Demand

Balancing of supply and demand is, of course, automatic. The issue is how to achieve this balance without causing unacceptable economic and social dislocations. Most of us old-time energy people almost automatically assumed that demand was hardly subject to any control: our emphasis, both as policy makers and as engineers, was on increasing supply. Our response to the oil embargo was to recommend the development of nuclear fission and fusion, oil shale, synfuels, geothermal, even solar and wind!

What a shock when we discovered that demand was also controllable—and, indeed, that the technologies required to reduce demand might challenge the engineering community no less than the technologies of expanding supply. Today, there are many opportunities for research in demand management as well as in supply enhancement.

Why did most engineers gravitate toward supply enhancement rather than demand management? I can give at least two reasons. First, designing nuclear reactors seemed to be more glamorous than improving efficiencies of cars. Second, demand management usually requires that millions of people change their way of doing something. In some cases, as lowering the thermostat, the change might affect lifestyle; in other cases, as replacing an energy-inefficient refrigerator with a more efficient one, it requires laying out additional money. In a broad sense, demand management, even when based on clever new technologies, is a social fix: lots of individual decisions are needed to achieve lower demand. By contrast, increasing supply was regarded, perhaps naively, as a purely technical fix: only a few people have to be convinced to build a nuclear reactor or a synfuel plant. The technical fix seemed to be simpler than the social fix.

At least, this is how many engineers viewed the matter. Somehow predicting the increased supply seemed much more robust than predicting reduced demand. But again we were wrong: we neglected the public antagonism to all sorts of large centralized energy systems—whether nuclear reactors, coal-fired power stations, or synfuel plants. And we underestimated the public's acceptance of conservation—

whether price induced, or resulting from a widespread belief in a conservation ethic reinforced by government-mandated efficiency standards.

The relative effectiveness of supply enhancement and demand management in a way reflects our underlying political structure. Ours is a Jeffersonian democracy: decentralized, open, some would say chaotic. Large-scale interventions that are perceived as threatening by a determined group can be, and often are, blocked. Under the circumstances, we have much incentive to avoid big, threatening energy-supply projects in favor of demand management and much smaller, decentralized supply options.

By contrast, where the political structure is elitist—particularly in France, with its Jacobin political tradition—large, centralized supply options, particularly nuclear energy, remain viable. France has managed to reduce oil imports (and, incidentally, to reduce the CO_2 it throws into the atmosphere) largely by its steadfast commitment to nuclear electrification—a path that is unavailable to us, at least for the present.

Incrementalism and Its Consequences

In trying to formulate energy strategy for the next decade, we must therefore accept three realities. The first is that the future is much less knowable than we thought it to be 15 years ago. The second is that ours has become a participatory polity, one with growing environmental concern. The third reality is that, although certain segments of our energy-supply system, notably oil, are dominated by a few large corporations, other segments, particularly electricity, are fragmented. The United States has about one generating company per million people, whereas Japan has one per 10 million and France has one per 50 million!

Our energy system seems to be responding to these realities with a grand strategy that I would describe as *incrementalism*. Since we cannot really predict what our energy demand will be in ten years, don't build anything that is very large. Power plants able to generate 1,000 megawatts or synfuel plants that can turn out 100,000 barrels per day are much too risky. If you need more electricity, build a 50- or 100-megawatt gas turbine, or buy electricity from small independent producers [who have enjoyed the protection of The Public Utilities Regulatory Policy Act (PURPA)], or, if possible, from Canada, and don't shut down old plants; or, reduce demand by offering incentives to customers to use more efficient devices. Such incrementalism finesses the uncertain future; and it does not expose a generating company to the risk of bankruptcy, which has engulfed unlucky utilities saddled with ridiculously overpriced nuclear reactors. Incrementalism also evokes little antagonism from politically sensitive, and often powerful, conservationists. Thus, the vision of the 1950s of a gradually nuclear electrified America has, in the 1980s and 1990s, given way to an America in which conservation is primary and in which energy is increasingly supplied by small, decentralized units and by older units that have been coaxed into a few more years of operation. The American energy dream of 1950 is coming to pass, but not in America. It is happening in France, in Japan, in several Communist countries, and in other newly industrialized countries whose political tradition is more authoritarian than ours and whose energy systems are more centralized.

Let us concede that incrementalism is inevitable during the next decade or so, at least if conservation is insufficient to keep our energy demand from growing. Do we see dangers in the long run for an energy system that is eventually dominated by a large number of small producers?

I see several such dangers. Perhaps most important, many of the new electrical supply increments use gas or oil-fired turbines. Although the Gas Research Institute (GRI) has recently estimated that there will be enough gas through 2010 (Table 2), one is naturally concerned that, to achieve GRI's projection of 19.9 quads of gas for 2010, some 8.4 quads must come from "new initiatives"—advanced technologies of extraction, the Alaskan pipeline, the Canadian Frontier, imports, and synthetics; and the total imports (including Canadian imports) amount to 3.6 quads. And, insofar as oil is used in these small generators, we will be increasing, not decreasing, our dependence on imported energy.

A second danger relates to economies of scale. During the era of energy euphoria, particularly nuclear euphoria, we assumed that bigger is cheaper. The catastrophic escalation of capital costs for nuclear plants has dissuaded us from this belief. We seem now to believe that the economic scaling laws have been repealed, or, at least, that they can be circumvented if the devices are manufactured serially. But even if the capital costs of small units are favorable, most would claim that operating costs will be higher for small plants than for large ones.

One must therefore ask, does the trend toward incrementalism imply that energy, particularly electrical energy, will always be more

TABLE 2. 1987 Gas Research Institute baseline projection of gas supplies.

	Gas supplies (quads)			
	Actual	Projected		
	1986	1990	2000	2010
Current Practice				
Domestic production	16.6	16.3	14.3	9.3
Canadian imports	0.8	0.9	1.4	1.1
LNG imports	0.0[a]	0.1	0.3	0.8
Supplemental sources	0.0[b]	0.2	0.3	0.3
	17.4	17.5	16.3	11.5
New Initiatives				
Lower-48 advanced technologies	0.0	0.3	2.6	5.4
Alaskan pipeline	0.0	0.0	0.0	1.2
Canadian Frontier	0.0	0.0	0.5	0.7
Other imports	0.0	0.0	0.0	1.0
Synthetics	0.0	0.0	0.0	0.1
	0.0	0.3	3.1	8.4
Total supply	17.4	17.8	19.4	19.9

[a]Less than 0.05 quad.
[b]Includes net injections to storage of 0.1 quad.

expensive in the United States than it is in such countries as France and Japan, not to speak of the Communist bloc, where large plants continue to dominate?

Though I am not optimistic about the next decade on this score, I see some hope for diminution in the price of energy in the trend toward extension of life of power plants and other large supply devices. The 30-year licensing lifetime of nuclear power plants was a relic of fossil-fueled tradition: fuel efficiency increased at a rate that made power from 30-year-old plants more expensive than from newly built plants. But with efficiencies plateauing—or being rather irrelevant, in the case of nuclear power plants—the incentive to shut down old plants has weakened. Extending the life of an old plant, whether nuclear or fossil-fuel, is often cheaper than building a new one. And if our energy system is dominated by plants that have already been paid off and have low operating and fuel costs, we may once more begin to see the price of energy fall.

I have speculated[11] that this phenomenon is not confined to standard electric generators but may also be applicable to synfuel plants or even to solar electric systems, always provided that operating costs are low. Thus, a synfuel plant, which might cost $100,000/daily barrel* ($330 per annual barrel) and uses coal at $40 per ton, will produce synfuel at about $92 per barrel; of this, $66 per barrel is capital cost (at 20 percent). But if the plant lasts, say, a century, rather than the 30 years over which it is amortized, and if its maintenance costs can be kept low, the cost of the synfuel falls to around $25 per barrel, once the plant has been amortized. The first SASOL plant in South Africa, which was placed in operation almost 30 years ago, and is now presumably amortized, probably produces synfuel at costs close to the world spot price of oil.

In constructing and modifying our energy system, we must recognize that we are dealing with one of our society's most basic infrastructures—and, like most infrastructures, it is built not only for our generation but for future generations as well. Perhaps the moral to be drawn is that future precepts of engineering design ought to stress longevity of the energy-producing device—more so than had been the case in the past. Though we may not succeed in giving our own generation the gift of cheap energy, perhaps we will be providing this gift to our children's children.

International Perspectives

My viewpoint has been primarily American; possibly, in view of my own background, it is also nuclear, electrical, and (despite my attempts to be neutral) supply-oriented. But I think it is fair to say that American energy policy during the coming decade will be demand-dominated—that the emphasis will be on trying to reduce demand. In this we are joined, for example, by Japan; the recent Ministry of International Trade and Industry (MITI) report[10] begins with an analysis of how well demand can be managed. But even MITI's minimum-demand scenario projects total energy in Japan expanding at 1.6

---

*Capital costs of a large production plant are reckoned in dollars per daily or yearly output. Thus, $100,000 per daily barrel means $100,000 to build a plant that would produce 1 barrel per day; $330 per annual barrel means $330 to build a plant that would produce one barrel per year.

TABLE 3. Forecasts of the world demand for energy (in millions of barrels per day of oil equivalents).[a]

		IIASA[b]		WEC[c]		Goldemberg *et al.*
Year	1980	2030		2020		2020
Primary energy demand		High	Low	High	Low	
World	145.4	497.0	310.7	348.8	271.1	158.2
Industrial nations	98.8	283.8	190.6	209.0	176.5	55.1
Developing countries	45.6	213.2	118.6	139.8	94.6	103.1

[a]Original values expressed in terawatts (TW) which were converted into million barrels/day (mmbd) equivalents. 1 TW = 14.12 mmbd oil equivalents.
[b]Analyses from the International Institute for Applied Systems Analysis, 1981.
[c]XII World Energy Conference, 1983.
Source: Goldemberg *et al.*, Annual Review of Energy **10**, p. 622 (1985).

percent per year until 2000 A.D. and at 0.8 percent per year from 2000 A.D. to 2030 A.D.—compared with a growth rate, from 1975 A.D. to 1985 A.D., of 1.3 percent per year. Thus, MITI expects the total energy demand in Japan to be at least 62 percent higher in 2030 A.D. than it is today; MITI's maximum-demand scenario projects that it will be 211 percent higher. Nor is Japan attracted to our American incrementalism: it expects to continue to build 1200-megawatt reactors, even as it diversifies supply.

As one looks at the entire world, one must be struck by the economic growth of the developing countries and by their increasing use of energy. But the diversity of estimates of the future remain (Table 3), from the World Energy Conference's high scenario of 735 quads in 2020 A.D. to Goldemberg's 320 quads in 2020 A.D. (about the same as today's 300 quads). Everyone seems to agree that the developing countries will use a larger fraction of the world's energy than they now do; but there is little agreement as to how much the total is going to be. Whether this presages a big spurt in oil prices—perhaps even a 1973- or 1979-style energy crisis during the next 20 years—or a much more gradual increase, no one can say.

Conclusions

Commonplace—even tiresome—is my observation that our forecasts were all wrong, that demand has ameliorated, and that we really don't know what is going to happen in the next decade or two. I have argued that, under such a veil of uncertainty, we can only try to choose policies, technological options, and strategies of research and development that are as resilient as possible to surprises. This suggests that the coming energy decade will be the decade of "creeping incrementalism": initiatives, whether enhancement of supply or diminution of demand, will be small and, in the short run, of low risk.

We have at least discovered what hasn't worked—in particular, that the magical talisman of nuclear energy simply was neither magical nor a talisman: it faltered because we nuclear optimists ignored social, political, and economic realities.

Perhaps what we have learned most of all is that the market is more powerful than government intervention. This was most strikingly demonstrated by the experience of the Synfuels Corporation. If the energy crisis was the moral equivalent of war, then why didn't the technique that worked so well in World War II, in the case of synthetic rubber, work for synfuels? Alas, markets can be circumvented by government ukase in wartime but not in peacetime. Synthetic oil was just too expensive!

So our experience with energy since 1973 seems to bear out the views of Friedrich Hayek more than those of John Maynard Keynes, let alone Karl Marx. During the 1930s, Hayek insisted that government intervention in the economy would inevitably fail because the detailed information on which the market operates could never be available to the government intervenors. (His debates with Keynes were high points in the history of economic thought.) Our experience of the past 15 years suggests that, on the whole, governmental interventions have not been very effective and that our fragmented, participatory political structure dooms government energy policies like those of France or Japan to failure here.

Does this mean that our best policy with respect to energy is to have no policy? Does it mean that we would do best to dismantle the Department of Energy, to stop all tax and direct subsidies, and to get the government out of energy policy—and even out of energy research and development?

I cannot accept such a conclusion. After all, some government interventions probably have helped—notably the CAFE standards, and the efficiency labeling of home appliances. And though markets are efficient, they are, as Mans Lonnroth points out, myopic; and they lack compassion. Although incrementalism (largely market-driven) seems inevitable at present, it may saddle us with unnecessarily expensive energy in the long run. Though nuclear energy is now too expensive and unpalatable to a large minority (if not a majority) of Americans, the incentive to develop inherently safe reactors that will be both economic and acceptable to the public remains strong. And, although doctrinal conservationists claim that demand management alone can defeat carbon dioxide, most engineers, I suspect, incline toward an eventual shift, worldwide, to nonfossil energy systems—most probably fission, but perhaps solar and fusion. Thus, a long-term role for government in providing the technical base for those elements of an energy system that are *not* mediated by the market seems to me to be proper.

We engineers are instinctively *technical* fixers; we are suspicious of social fixes, perhaps because we don't understand them, perhaps because we regard them as more difficult than technical fixes. As we look to an unknowable future, we recognize that, despite its social components, our energy future will depend ultimately on *technical* ingenuity, on cheap variable-speed motors, on practical energy-storage devices, on more efficient cars, on economic photovoltaic systems, on inherently safe reactors—perhaps even on successful fusion. The menu of technical challenges is large, much larger than we realized when the Joint Committee on Atomic Energy dominated our government energy research-and-development policy. A challenge so large and so diverse is what we engineers thrive upon.

References

[1] P. Putnam, *Energy in the Future* (Van Nostrand, New York, 1953).

[2] President's Material Policy Commission, *Resources for Freedom, A Report to the President* (U.S. GPO, Washington, DC, 1952).

[3] M. K. Hubbert, "Energy Resources" in Committee on Resources and Man of National Academy of Sciences *Resources and Man*, Freeman, San Francisco (1969).

[4] A. Cambel, *Energy R&D and National Progress*, Interdepartmental Energy Study Group (U.S. GPO, Washington, DC, 1964).

[5] D. L. Ray, *The Nation's Energy Future, A Report to Richard M. Nixon, President of the United States* (U.S. GPO, Washington, DC, 1973).

[6]Federal Energy Administration, *Project Independence, A Summary* (U.S. GPO, Washington, DC, 1974).

[7]A. Lovins, "Energy Strategy, The Road Not Taken?," in *Future Strategies for Energy Development, A Question of Scale* (Oak Ridge Associated Universities, Oak Ridge, TN, 1977).

[8]Energy Policy Project, *A Time to Choose, America's Energy Future* (Ballinger Press, Cambridge, MA, 1974).

[9]H. Brooks and J. M. Hollander, "United States Energy Alternatives to 2010 and Beyond: The CONAES Study," in *Annual Review of Energy 4*, pp. 1–70 (Annual Reviews, Inc., Palo Alto, CA, 1979).

[10]MITI (no date). *The Twenty-First Century Energy Vision—Entering the Multiple Energy Era*, Ministry of International Trade & Industry, Japan.

[11]A. M. Weinberg, "Immortal Energy Systems & Intergenerational Justice," *Energy Policy* **13** (1), 51–59 (1985).

Part V: Nuclear Energy

My technical career has coincided with the First Nuclear Era: the years from 1939 when fission was discovered, to today, 1991, when nuclear energy has been rejected by most of the countries, including the United States, that have deployed nuclear power reactors. Most of my publications during these fifty years have dealt with nuclear energy. Like all of us in the nuclear enterprise, I was a strong nuclear optimist—nuclear energy would indeed, as H. G. Wells said, "set the world free." This view permeated my first book of essays, *Reflections on Big Science*, which was published in 1967. But by 1970, doubt crept into my mind. This transition was captured in my book *Continuing the Nuclear Dialogue* published in 1985.

The essays in this section, spanning the period from 1972 to 1991, are concerned mostly with how to fix nuclear energy—what must be done to allow a rebirth of nuclear energy, "a second nuclear era." They reflect the views we developed at the Institute for Energy Analysis of the Oak Ridge Associated Universities. The Institute was formed in 1974, and I directed it from 1975 to 1985.

"*Social Institutions and Nuclear Energy,*" which I wrote in 1972, calls nuclear energy a "Faustian bargain." This characterization has often been used by nuclear opponents as a justification for their disapproval of nuclear energy; and many of my colleagues in the nuclear enterprise have scolded me for providing such ammunition to our opponents. I, however, still think the phrase Faustian bargain is justified, especially since Goethe's Faust was eventually redeemed.

I wrote "Salvaging the Atomic Age" in the heat of the accident at Three Mile Island. A few years earlier, our Institute for Energy Analysis had warned that an accident like the one at Three Mile Island was likely and that, even if no one was hurt, the accident could profoundly affect the future of nuclear energy. The recipes for fixing nuclear energy that I proposed in this article had been worked out at the Institute between 1975, when the Institute was founded, and 1979, when the accident at Three Mile Island happened.

"Burning the Rocks Forty Years Later" was presented in Chicago at the twenty-fifth anniversary celebration of Walter Zinn's prototype breeder, EBR-II. I had coined the expression "Burning the Rocks" in a paper "Energy as an Ultimate Raw Material," published in 1959 in *Physics Today*. There I pointed out that *both* thermonuclear reactors that "burn the sea," and breeder reactors that "burn the rocks," were inexhaustible sources of energy. This realization explains my conviction that breeder development is essential—but, as things have turned out, only if engineers can bring the capital cost of the breeder into line with the cost of other energy sources.

"Nuclear Power and Public Perception" was presented in 1987 at a seminar chaired by Professor Martin Shubik, the Knox Professor of Mathematical Institutional Economics at Yale. I used this occasion to argue that the acceptability of nuclear power depended heavily on the social ethos on which a society's political structure is based.

"Engineering in an Age of Anxiety" was given at the twenty-fifth anniversary of the founding of the National Academy of Engineering. Our anxieties about being harmed by environmental threats seem to be worse in 1991 than they were in 1989 when I gave this paper.

"Nuclear Energy and the Greenhouse Effect" was given in 1989 at the Midwest Universities Energy Consortium Symposium on Nuclear Safety. Our Institute for Energy Analysis was the first American think tank to examine the greenhouse effect from both a technical and a societal standpoint. My main contribution to the greenhouse issue was to call the attention of Washington's Research and Development Administrators to the steady yearly increase of carbon dioxide in the atmosphere measured by Dave Keeling at Mauna Loa. I remember going from office to office in Washington, carrying Keeling's carbon dioxide curves in hand; and at the same time pointing out that the prime non-carbon dioxide producing energy source, nuclear fission, was all but dead (this was in 1975). This paper summarizes my reasons for fixing nuclear energy, not extirpating it.

I presented "The First and Second Fifty Years of Nuclear Fission" in 1991 at an Oak Ridge conference on Technologies for a Greenhouse-Constrained Society. I had been asked to give a historical overview of nuclear energy. Since I am now writing my own memoirs,

I was able to reconstruct events that happened 50 years ago. As I think about those days, I marvel at how naive we were. Only Fermi had the foresight to realize that nuclear energy might not be accepted by the public, even though it seemed to us to be the ideal energy source!

Social Institutions and Nuclear Energy

Fifty-two years have passed since Ernest Rutherford observed the nuclear disintegration of nitrogen when it was bombarded with alpha particles. This was the beginning of modern nuclear physics. In its wake came speculation about the possibility of releasing nuclear energy on a large scale: By 1921, Rutherford was saying "The race may date its development from the day of the discovery of a method of utilizing atomic energy."[1]

Despite the advances in nuclear physics beginning with the discovery of the neutron by Chadwick in 1932 and Cockcroft and Walton's method for electrically accelerating charged particles, Rutherford later became a pessimist about nuclear energy. Addressing the British Association for the Advancement of Science in 1933, he said: "We cannot control atomic energy to an extent which would be of any value commercially, and I believe we are not likely ever to be able to do so."[2] Yet Rutherford did recognize the great significance of the neutron in this connection. In 1936, after Fermi's remarkable experiments with slow neutrons, Rutherford wrote ". . . the recent discovery of the neutron and the proof of its extraordinary effectiveness in producing transmutations at very low velocities opens up new possibilities, if only a method could be found of producing slow neutrons in quantity with little expenditure of energy."[3]

Based on the text of the Rutherford Centennial lecture, presented at the annual meeting of the American Association for the Advancement of Science, Philadelphia, 1971. First published in *Science* **177**, 27–34, 1972. Reprinted by permission.

Today, the United States is committed to more than 100×10^6 kilowatts of nuclear power, and the rest of the world to an equal amount. Rather plausible estimates suggest that by 2000 A.D. the United States may be generating electricity at a rate of $1{,}000 \times 10^6$ kilowatts with nuclear reactors. Much more speculative estimates visualize an ultimate world of 15 billion people, living at something like the current U.S. standard: nuclear fission might then generate power at the rate of some 300×10^9 kilowatts of heat, which represents 1/400 of the flux of solar energy absorbed and reradiated by the Earth.[4]

This large commitment to nuclear energy has forced many of us in the nuclear community to ask, with the utmost seriousness, questions that, when first raised, had a tone of unreality. When nuclear energy was small and experimental and unimportant, the intricate moral and institutional demands of a full commitment to it could be ignored or not taken seriously. Now that nuclear energy is on the verge of becoming our dominant form of energy, questions about the adequacy of human institutions to deal with this marvelous new kind of fire must be asked, and answered, soberly and responsibly. In these remarks, I review in broadest outline where the nuclear energy enterprise stands and what I think are its most troublesome problems; and I shall then speculate on some of the new and peculiar demands our commitment to nuclear energy may impose on human institutions.

Nuclear Burners—Catalytic and Noncatalytic

Even before Fermi's experiment at Stagg Field on 2 December 1942, reactor design had captured the imagination of many physicists, chemists, and engineers at the Chicago Metallurgical Laboratory. Almost without exception, each of the two dozen main reactor types developed during the following 30 years had been discussed and argued over during those frenzied war years. Of these various reactor types, about five—moderated by light water, heavy water, or graphite—have survived. In addition, breeders, most notably the sodium-cooled plutonium breeder, are now under active development.

Today, the dominant reactor type uses enriched uranium oxide fuel and is moderated and cooled by water at pressures of 100 to 200 atmospheres. The water may generate steam directly in the reactor (the so-called boiling-water reactor, BWR) or it may transfer its heat to an external steam generator (the pressurized-water reactor, PWR). These light-water reactors (LWR) require enriched uranium and therefore at first could be built only in such countries as the United States and the U.S.S.R., which had large plants for separating uranium isotopes.

In countries where enriched uranium was unavailable or was much more expensive than in the United States, reactor development went along directions that utilized natural uranium: for example, reactors developed in the United Kingdom and France were based mostly on the use of graphite as a moderator; those developed in Canada used D_2O (deuterium oxide, or heavy water) as a moderator. Both D_2O and graphite absorb fewer neutrons than H_2O, and therefore such reactors can be fueled with natural uranium. However, as enriched uranium has become more generally available (of the uranium above ground, probably more by now has had its normal isotopic ratio altered than not), the importance of the natural ^{235}U isotopic abundance of 0.71 percent has faded. All reactor systems now tend to use at least slightly enriched uranium, since its use gives the designer more leeway with respect to materials of construction and configuration of the reactor.

The PWR was developed originally for submarine propulsion, where compactness and simplicity were the overriding considerations. As one who was closely involved in the very early thinking about the use of pressurized water for submarine propulsion (I still remember the spirited discussions we used to have in 1946 with Captain Rickover at Oak Ridge over the advantages of the pressurized-water system), I am still a bit surprised at the enormous vogue of this reactor type for civilian power. Compact, and in a sense simple, these reactors were; but in the early days we hardly imagined that separated ^{235}U would ever be cheap enough to make such reactors really economical as sources of central station power.

Four developments proved us wrong. First, separated ^{235}U, which at the time of *Nautilus* cost around $100 per gram, fell to $12 per gram. Second, the price of coal rose from around $5 per ton to $8 per ton. Third, oxide fuel elements, which use slightly enriched fuel rather than the highly enriched fuel of the original LWR, were developed. This meant that the cost of fuel in an LWR could be, say, 1.9 mills per kilowatt hour (compared with around 3 mills per electric kilowatt hour for a coal-burning plant with coal at $8 per ton). Fourth, pressure vessels of a size that no one could have envisioned in 1946 were common by 1970: the pressure vessel for a large PWR may be as much as $8\frac{1}{2}$ inches thick and 44 feet tall. Development of these large pressure vessels made possible reactors of 1000 megawatts electric (MWe) or more, compared with 60 MWe at the original Shippingport reactor. Since, per unit of output, a large power plant is cheaper than a small one, this increase in reactor size was largely responsible for the economic breakthrough of nuclear power.

Water-moderated reactors burn ^{235}U, which is the only naturally occurring fissile isotope. But the full promise of nuclear fission will be achieved only with successful breeders. These are reactors that, essentially, burn the very abundant isotopes ^{238}U or ^{232}Th; in the process, fissile ^{239}Pu or ^{233}U acts as regenerating catalyst—that is, these isotopes are burned and regenerated. I therefore like to call reactors of this type *catalytic nuclear burners*. Since ^{238}U and ^{232}Th are immensely abundant (though in dilute form) in the granitic rocks, the basic fuel for such catalytic nuclear burners is, for all practical purposes, inexhaustible. Mankind will have a permanent source of energy, once such catalytic nuclear burners are developed.

Most of the world's development of a breeder is centered around the sodium-cooled, ^{238}U burner, in which ^{239}Pu is the catalyst and in which the energy of the neutrons is above 100×10^3 electron volts. No fewer than 12 reactors of this liquid-metal fast-breeder reactor (LMFBR) type are being worked on actively, and the United Kingdom plans to start a commercial 1000-MWe fast breeder by 1975. Some work continues on alternatives: in the ^{233}U–^{232}Th cycle, on the light-water breeder and the molten-salt reactor; in the ^{239}Pu–^{238}U cycle, on the gas-cooled fast breeder. But these systems are, at least at the present, viewed as backups for the main line, which is the LMFBR.

Nuclear Power and Environment

The great surge to nuclear power is easy to understand. In the short run, nuclear power is cheaper than coal power in most parts of the United States; in the long run, nuclear breeders assure us of an all but inexhaustible source of energy. Moreover, a *properly* operating nuclear power plant and its subsystems (including transport, waste disposal, chemical plants, and even mining) are, except for the heat load, far less damaging to the environment than a coal-fired plant would be.

The most important emissions from a routinely operating reactor are heat and a trace of radioactivity. Heat emissions can be summarized quickly. The thermal efficiency of a PWR is 32 percent; that of a modern coal-fired power plant is around 40 percent. For the same electrical output, the nuclear plant emits about 40 percent more waste heat than the coal plant does; in this one respect, present-day nuclear plants are more polluting than coal-fired plants. However, the higher-temperature nuclear plants, such as the gas-cooled, the molten-salt breeder, and the liquid-metal fast breeder, operate at about the same efficiency as does a modern coal-fired plant. Thus, nuclear reactors of the future ought to emit no more heat than other sources of thermal energy.

As for routine emission of radioactivity, even when the allowable maximum exposure to an individual at the plant boundary was set at 500 millirems (mrem) per year, the hazard, if any, was extremely small. But, for practical purposes, technological advances have all but eliminated routine radioactive emission. These improvements are taken into account in the newly proposed regulations of the Atomic Energy Commission (AEC), which require, in effect, that the dose imposed on any individual living near the plant boundary, either by liquid or by gaseous effluents from LWR's, should not exceed 5 mrem per year. This is to be compared with the natural background, which is around 100 to 200 mrem per year, depending on location, or the medical dose, which now averages around 60 mrem per year.

As for emissions from chemical reprocessing plants, data are relatively scant, since but one commercial plant, the Nuclear Services Plant at West Valley, New York, has been operating, and this only since 1966. During this time, liquid discharges have imposed an average dose of 75 mrem per year at the boundary. Essentially no ^{131}I has been emitted. As for the other main gaseous effluents, all the ^{85}Kr and ^{3}H contained in the fuel has been released. This has amounted to an average dose from gaseous discharge of about 50 mrem per year.

Technology is now available for reducing liquid discharges, and processes for retaining ^{85}Kr and ^{3}H are being developed at AEC laboratories. There is every reason to expect these processes to be successful. Properly operating radiochemical plants in the future should emit no more radioactivity than do properly operating reactors—that is, less than 10 percent of the natural background at the plant boundary.

There are some who maintain that even 5 mrem per year represents an unreasonable hazard. Obviously, there is no way to decide whether there is any hazard at this level. At this stage, the argument passes from science into the realm of what I call trans-science, and one can only leave it at that.

My main point is that nuclear plants are indeed relatively innocuous, large-scale power generators, if they and their subsystems work properly. The entire controversy that now surrounds the whole nuclear-power enterprise therefore hangs on the answer to the question of whether nuclear systems can be made to work properly—or, if faults develop, whether the various safety systems can be relied upon to guarantee that no harm will befall the public.

The question has only one answer: there is no way to guarantee that a nuclear reactor and all of its subsystems will never cause harm. But I shall try to show why I believe the measures that have been taken, and are being taken, have reduced to an acceptably low level the probability of damage.

I have already discussed low-level radiation and the thermal emissions from nuclear systems. Of the remaining possible causes of concern, I shall dwell on the three that I regard as most important: reactor safety, transport of radioactive materials, and permanent disposal of radioactive wastes.

Avoiding Large Reactor Accidents

One cannot say categorically that a catastrophic failure of a large PWR or a BWR and its containment is impossible. The most elaborate measures are taken to make the probability of such occurrence extremely small. One of the prime jobs of the nuclear community is to consider all events that could lead to accident and, by proper design, to keep reducing their probability, however small it may be. On the other hand, there is some danger that, in mentioning the matter, one's remarks may be misinterpreted as implying that the event is likely to occur.

Assessment of the safety of reactors depends upon two rather separate considerations: prevention of the initiating incident that would require emergency safety measures; and assurance that the emergency measures, such as the emergency core cooling, if ever called upon, would work as planned. In much of the discussion and controversy that has been generated over the safety of nuclear reactors, emphasis has been placed on what would happen if the emergency measures were called upon and failed to work. But, to most of us in the reactor community, this is secondary to another question: How certain can we be that a drastic accident that calls into play the emergency systems will never happen? What one is counting upon for the safety of a reactor is primarily the integrity of the primary cooling system—that

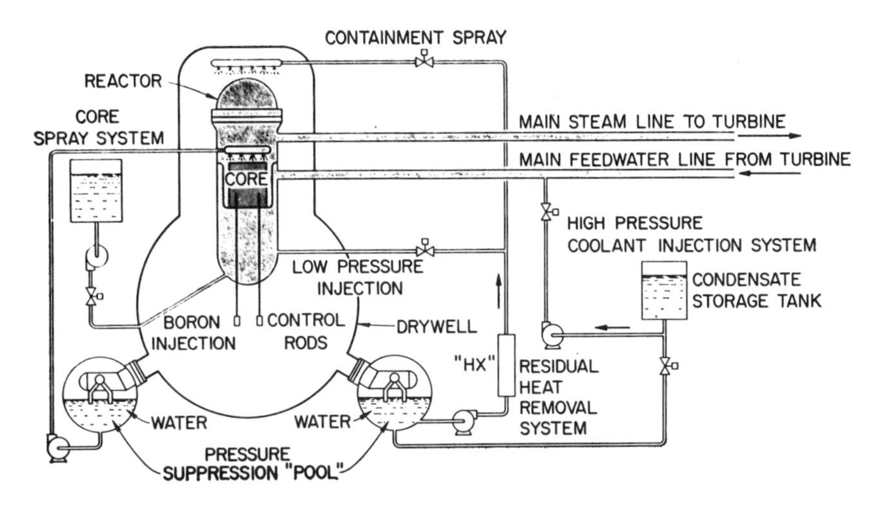

FIGURE 1. *Boiling-water reaction, showing emergency cooling systems.*

is, the integrity of the pressure vessel and the pressure piping. Excruciating pains are taken to assure the integrity of these vessels and pipes. The watchword throughout the nuclear reactor industry is *quality assurance*: every piece of hardware in the primary system is examined and reexamined to guarantee, insofar as possible, that there are no flaws.

Nevertheless, we must deal with the remote contingency that might call the emergency systems into action. How certain can one be that these will work as planned? To understand better the analysis of the emergency system, Figures 1 and 2 show, schematically, a large BWR and a PWR.

Three barriers prevent radioactivity from being released: fuel-element cladding, primary pressure system, and containment shell. In addition to the regular safety system, consisting primarily of the control and safety rods, there are elaborate provisions for preventing the residual radioactive heat from melting the fuel in the event of a loss of coolant. In the BWR, there are sprays that spring into action within 30 seconds of an accident. In both the PWR and BWR, water is injected under pressure from gas-pressurized accumulators. In both reactors, there are additional systems for circulating water after the system has come to low pressure, as well as means for reducing the pressure of

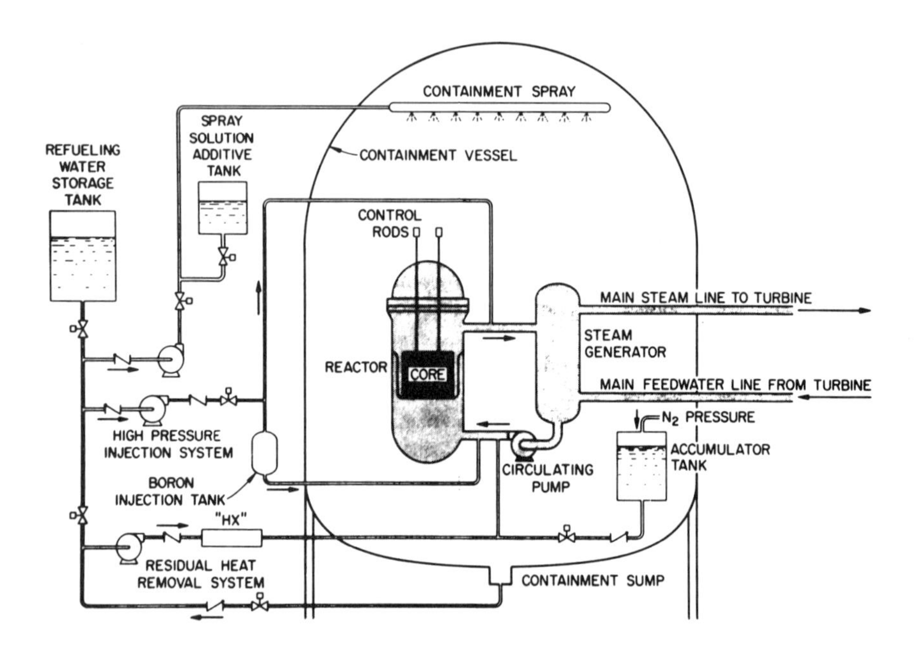

FIGURE 2. *Pressurized-water reactor, showing emergency cooling systems.*

steam in the containment vessel. This latter system also washes down, or otherwise helps remove, any fission products that may become airborne.

In analyzing the ultimate safety of a LWR, one tries to construct scenarios—improbable as they may be—of how a catastrophe might occur, and then one tries to provide reliable countermeasures for each step in the chain of failures that could lead to catastrophe. The chain, conceivably, could go like this. First, a pipe might break, or the safety system might fail to respond when called upon in an emergency. Second, the emergency core-cooling system might fail. Third, the fuel might melt, might react also with the water, and, conceivably, might melt through the containment. Fourth, the containment might fail catastrophically—if not from the melt itself, then from missiles or overpressurization—and radioactivity might then spread to the public. There may be other modes of catastrophic failure—for example, earthquakes or acts of violence—but the preceding sequence is the one most commonly envisioned.

To give the flavor of how the analysis of an accident is made, let me say a few words about the first and second steps of this chain. As a first step, one might imagine a failure of the safety system to respond in an emergency—say, when the bubbles in a BWR collapse after a fairly routine turbine trip. Here the question is not whether some safety rods will work and some will not but, rather, whether a common mode failure might render the entire safety system inoperable. Thus, if all the electrical cables actuating the safety rods were damaged by fire, this would be a common mode failure. Such a common mode failure is generally regarded as impossible, since the actuating cables are carefully segregated, as are groups of safety rods, so as to avoid such an accident. But one cannot *prove* that a common mode failure is impossible. It is noteworthy that, on 30 September 1970, the entire safety system of the Hanford-N reactor (a one-of-a-kind water-cooled, graphite-moderated reactor) did fail when called upon; however, the backup samarium balls dropped precisely as planned and shut off the reactor. One goes a long way toward making such a failure incredible if each big reactor, as in the case of the Hanford-N reactor, has two entirely independent safety systems that work on totally different principles. In the case of BWR, shutoff of the recirculation pumps in the all but incredible event the rods fail to drop constitutes an independent shutoff mechanism, and automatic pump shutoff is being incorporated in the design of modern BWRs.

The other step in the chain that I shall discuss is the failure of the emergency core-cooling system. At the moment, there is some controversy whether the initial surge of emergency core-cooling water would bypass the reactor or would in fact cool it. The issue was raised recently by experiments on a very small scale (9-inch-diameter pot), which indeed suggested that the water in that case would bypass the core as the pressure in the reactor is relieved as the result of an inadvertent loss of coolant. However, there is a fair body of experts within the reactor community who hold that these experiments were not sufficiently accurate simulations of an actual PWR to bear on the reliability or lack of reliability of emergency core cooling in a large reactor.

Obviously, the events following a catastrophic loss of coolant and injections of emergency coolant are complex. For example, one must ask whether the fuel rods will balloon and block coolant channels, whether significant chemical reactions will take place, or whether the fuel cladding will crumble and allow radioactive fuel pellets to fall out.

Such complex sequences are hardly susceptible to a complete analysis. We shall never be able to estimate everything that will happen in a loss-of-coolant accident with the same kind of certainty with which we can compute the Balmer series or even the course of the ammonia synthesis reaction in a fertilizer plant. The best that we can do as knowledgeable and concerned technologists is to present the evidence we have, and to expect policy to be based upon informed—not uninformed—opinion.

Faced with questions of this weight, which in a most basic sense are not fully susceptible to a *yes* or *no* scientific answer, the AEC has invoked the adjudicatory process. The issue of the reliability of the emergency core-cooling system is being taken up in hearings before a special board drawn from the Atomic Safety and Licensing Board Panel. The record of the hearings is expected to contain all that is known about emergency core-cooling systems and to provide the basis for setting the criteria for design of such systems.

Transport of Radioactive Materials

If, by the year 2000, we have 10^6 megawatts of nuclear power, of which two-thirds are liquid metal fast breeders, then there will be 7,000 to 12,000 annual shipments of spent fuel from reactors to chemical plants, with an average of 60 to 100 loaded casks in transit at all times. Projected shipments might contain 1.5 tons of core fuel that has decayed for as little as 30 days, in which case each shipment would generate 300 kilowatts of thermal power and 75 megacuries of radioactivity. By comparison, present casks from LWRs might produce 30 kilowatts and contain 7 megacuries.

Can we estimate the hazard associated with transport of these materials? The derailment rate in rail transport (in the United States) is 10^{-6} per car mile. Thus, if there were 12,000 shipments per year, each of a distance of 1000 miles, we would expect 12 derailments annually. However, the number of serious accidents would be perhaps 10^{-4} to 10^{-6} times less frequent; and shipping casks are designed to withstand all but the most serious accident (the train wreck near an oil refinery that goes into flames as a result of the crash). Thus the statistics—between 1.2×10^{-3} and 1.2×10^{-5} serious accidents per year—at least until the year 2000, look quite good. Nevertheless, the shipping problem is a difficult one and may force a change in basic strategy. For example, we may decide to cool fuel from LMFBRs in place for 360 days before shipping: this reduces the heat load sixfold, and increases the cost of power by only around 0.2 mill per electric kilowatt hour. A solution that I personally prefer is to cluster fast breeders in nuclear-power parks, which would have their own on-site reprocessing facilities.[5] Clustering reactors in this way would make both cooling and transmission of power difficult; also such parks would be more vulnerable to common-mode failure, such as acts of war or earthquakes. These difficulties must be balanced against the advantage of not shipping spent fuel elsewhere for reprocessing and of simplifying control of fissile material against diversion. To my mind, the advantages of clustering outweigh its disadvantages; but this again is a trans-scientific question that can be settled only by a legal or political process, rather than by scientific exchange among peers.

Waste Disposal

By the year 2000, according to present (1971) projections, we shall have to sequester about 27,000 megacuries of radioactive wastes in the United States; these wastes will be generating 100,000 kilowatts of heat at that time.

The wastes will include about 400 megacuries of transuranic alpha emitters. Of these, the ^{239}Pu with a half-life of 24,400 years will be dangerous for perhaps 200,000 years.

Can we see a way of dealing with these unprecedentedly treacherous materials? I believe we can, but not without complication.

There are two basically different approaches to handling the wastes. The first, urged by W. Bennett Lewis of Chalk River,[6] argues that once man has opted for nuclear power he has committed himself to essentially perpetual surveillance of the apparatus of nuclear power, such as the reactors, the chemical plants, and others. Therefore, so the argument goes, there will be spots on the Earth where radioactive operations will be continued in perpetuity. The wastes then would be stored at these spots, say in concrete vaults. Lewis further refines his ideas by suggesting that the wastes be recycled so as to limit their volume. As fission products decay, they are removed and thrown away as innocuous nonradioactive species; the transuranics are sent back to the reactors to be burned. The essence of the scheme is to keep the wastes under perpetual, active surveillance and even processing. This is deemed possible because the original commitment to nuclear energy is considered to be a commitment in perpetuity.

There is merit in these ideas; and indeed permanent storage in vaults is a valid proposal. However, if one wishes to perpetually rework the wastes as Lewis suggests, chemical separations would be required that are much sharper than those we now know how to do; otherwise at every stage in the recycling we would be creating additional low-level wastes. We probably can eventually develop such sharp separation methods; but these, at least with currently visualized techniques, would be very expensive. It is on this account that I like better the other approach, which is to find some spot in the universe where the wastes can be placed forever out of contact with the biosphere. Now the only place where we know absolutely the wastes will never interact with the biosphere is in far outer space. But the roughly estimated cost of sending wastes into permanent orbit with foreseeable rocket technology is in the range of 0.2 to 2 mills per electric kilowatt hour, not to speak of the hazard of an abortive launch. For both these reasons I do not count on rocketing the wastes into space.

This pretty much leaves us with disposal in geologic strata. Of the many possibilities—deep rock caverns, deep wells, bedded salt—the latter has been chosen, at least on an experimental basis, by the United States and West Germany. The main advantages of bedded salt are primarily that, because salt dissolves in water, the existence of a stratum of bedded salt is evidence that the salt has not been in contact with circulating water during geologic time. Moreover, salt flows plastically; if radioactive wastes are placed in the salt, eventually the salt ought to envelop the wastes and sequester them completely.

These arguments were adduced by the National Academy of Sciences Committee on Radioactive Waste Management[7] in recommending that the United States investigate bedded salt (which underlies 500,000 square miles in our country) for permanent disposal of radioactive wastes. And, after 15 years of discussion and research, the AEC in 1970 decided to try large-scale waste disposal in an abandoned salt mine in Lyons, Kansas. If all goes as planned, the Kansas mine is to be used until 2000 A.D. What one does after 2000 A.D. would of course depend on our experience during the next 30 years (1970 to 2000). In any event, the mine is to be designed so as to allow the wastes to be retrieved during this time.

The salt mine is 1,000 feet deep, and the salt beds are around 300 feet thick. The beds were laid down in Permian times and had been undisturbed, until man himself intruded, for 200 million years. Experiments in which radioactive fuel elements were placed in the salt have clarified details of the temperature distribution around the wastes, the effect of radiation on salt, the migration of water of crystallization within the salt, and so on.

The general plan is first to calcine the liquid wastes to a dry solid. The solid is then placed in metal cans, and the cans are buried in the floor of a gallery excavated in the salt mine. After the floor of the gallery is filled with wastes, the gallery is backfilled with loose salt.

Eventually this loose salt will consolidate under the pressure of the overburden, and the entire mine will be resealed. The wastes will have been sequestered, one hopes, forever.

Much discussion has centered around the question of just how certain we are that the events will happen exactly as we predict. For example, is it possible that the mine will cave in and that this will crack the very thick layers of shale lying between the mine and an aquifer at 200 feet below the surface? There is evidence to suggest that this will not happen, and I believe most, though not all, geologists who have studied the matter agree that the 500-foot-thick layer of shale above the salt is too strong to crack so completely that water could enter the mine from above.

But human interventions are not so easily disposed of. In Kansas there are some 100,000 oil wells and dry holes that have been drilled through these salt formations. These holes penetrate aquifers; in principle, they can let water into the mine. For the salt mine to be acceptable, one must plug all such holes. At the originally proposed site, there were 30 such holes; in addition, solution mining was practiced nearby. For this reason, the AEC recently authorized the Kansas State Geological Survey to study other sites that were not peppered with man-made holes. The AEC also announced recently its intention to store solidified wastes in concrete vaults, pending resolution of these questions concerning permanent disposal in geologic formations.

Human intervention complicates the use of salt for waste disposal; yet by no means does this imply that we must give up the idea of using salt. In the first place, such holes can be plugged, though this is costly and requires development. In the second place, let us assume the all but incredible event that the mine is flooded—let us say 10,000 years hence. By that time, since no new waste will have been placed in the mine after 2000 A.D., all the highly radioactive beta-decaying species, notably ^{90}Sr and ^{137}Cs, will have decayed. The main radioactivity would then come from the alpha emitters. The mine would contain 38 tons of ^{239}Pu mixed with about a million tons of nonradioactive material. The plutonium in the cans is thus diluted to 38 parts per million; since plutonium is, per gram, 10,000 times more hazardous than natural uranium in equilibrium with its daughters, these diluted waste materials would present a hazard of the same order as an equal amount of pitchblende. Actually, the 38 tons of ^{239}Pu is spread over 200 acres. If all the salt associated with the ^{239}Pu were dissolved in water, as conceivably could result from total flooding of the mine, the concentration of plutonium in the resulting salt solution would be well below maximum permissible concentrations. In other words, by virtue of having spread the plutonium over an area of 200 acres, we will have, to a degree, ameliorated the residual risk in the most unlikely event that the mines are flooded.

Despite such assurances, the mines must not be allowed to flood, especially before the ^{137}Cs and ^{90}Sr decay. We must prevent human intrusion—and this can be assured only by humans themselves. Thus we again come back to the great desirability, if not absolute necessity in this case, of keeping the wastes under some kind of surveillance in perpetuity. The great advantage of the salt method over, say, the perpetual reworking method, or even the aboveground concrete vaults without reworking, is that our commitment to surveillance in the case of salt is minimal. All we have to do is prevent people from intruding, rather than keeping a priesthood that forever reworks the wastes or guards the vaults. And if the civilization should falter, which would mean, among other things, that we abandon nuclear power altogether, we can be almost (but not totally) assured that no harm would befall our recidivist descendants of the distant future.

Social Institutions—Nuclear Energy

We nuclear people have made a Faustian bargain with society. On the one hand, we offer—in the catalytic nuclear burner—an inexhaustible source of energy. Even in the short range, when we use ordinary reactors, we offer energy that is cheaper than energy from fossil fuel. Moreover, this source of energy, when properly handled, is almost nonpolluting. Whereas fossil fuel burners must emit oxides of carbon and nitrogen, and probably will always emit some sulfur dioxide, there is no intrinsic reason why nuclear systems must emit any pollutant—except heat and traces of radioactivity.

But the price that we demand of society for this magical energy source is both a vigilance and a longevity of our social institutions that we are quite unaccustomed to. In a way, all of this was anticipated during the old debates over nuclear weapons. As matters have turned out, nuclear weapons have stabilized at least the relations between the superpowers. The prospects of an all-out third world war seem to recede. In exchange for this atomic peace we have had to manage and control nuclear weapons. In a sense, we have established a military priesthood which guards against inadvertent use of nuclear weapons, which maintains what *a priori* seems to be a precarious balance between readiness to go to war and vigilance against human errors that would precipitate war. Moreover, this is not something that will go away, at least not soon. The discovery of the bomb has imposed an additional demand on our social institutions. It has called forth this military priesthood upon which in a way we all depend for our survival.

It seems to me (and in this I repeat some views expressed very well by Atomic Energy Commissioner Wilfrid Johnson) that peaceful nuclear energy probably will make demands of the same sort on our society, and possibly of even longer duration. To be sure, we shall steadily improve the technology of nuclear energy; but, short of developing a truly successful thermonuclear reactor, we shall never be totally free of concern over reactor safety, transport of radioactive materials, and waste disposal. And even if thermonuclear energy proves to be successful, we shall still have to handle a good deal of radioactivity.

We make two demands. The first, which I think is the easier to manage, is that we exercise in nuclear technology the very best techniques and that we use people of high expertise and purpose. Quality assurance is the phrase that permeates much of the nuclear community these days. It connotes using the highest standards of engineering design and execution; of maintaining proper discipline in the operation of nuclear plants in the face of the natural tendency to relax as a plant becomes older and more familiar; and perhaps of managing and operating our nuclear power plants with people of higher qualification than were necessary for managing and operating nonnuclear power plants: in short, of creating a continuing tradition of meticulous attention to detail.

The second demand is less clear, and I hope it may prove to be unnecessary. This is the demand for longevity in human institutions. We have relatively little problem dealing with wastes if we can assume always that there will be intelligent people around to cope with eventualities we have not thought of. If the nuclear parks that I mention are permanent features of our civilization, then we presumably have the social apparatus, and possibly the sites, for dealing with our wastes indefinitely. But even our salt mine may require some small measure of surveillance if only to prevent people in the future from drilling holes into the burial grounds.

Eugene Wigner has drawn an analogy between this commitment to a permanent social order that may be implied in nuclear energy and our commitment to a stable, year-in and year-out social order when we moved from hunting and gathering to agriculture. Before agriculture, social institutions hardly required the long-lived stability that we now take so much for granted. And the commitment imposed by agriculture in a sense was forever: the land had to be tilled and irrigated every year in perpetuity; the expertise required to accomplish this task could not be allowed to perish or we would perish; our numbers could not be sustained by hunting and gathering. In the same sense, though on a much more highly sophisticated plane, the knowledge and care that goes into the proper building and operation of nuclear power plants and their subsystems is something that we are committed to forever, so long as we find no other practical energy source of infinite extent.[8]

Let me close on a somewhat different note. The issues I have discussed here—reactor safety, waste disposal, transport of radioactive materials—are complex matters about which little can be said with absolute certainty. When we say that the probability of a serious reactor incident is perhaps 10^{-8} or even 10^{-4} per reactor per year, or that the failure of all safety rods simultaneously is incredible, we are speaking of matters that simply do not admit of the same order of scientific certainty as when we say it is incredible for heat to flow against a temperature gradient or for a perpetuum mobile to be built. As I have said earlier, these matters have trans-scientific elements. We claim to be responsible technologists, and as responsible technologists we give as our judgment that these probabilities are extremely—almost vanishingly—small; but we can never represent these things as certainties. The society must then make the choice, and this is a choice that we nuclear people cannot dictate. We can only participate in making it. Is mankind prepared to exert the eternal vigilance needed to ensure proper and safe operation of its nuclear energy system? This admittedly is a significant commitment that we ask of society. What we offer in return, an all but infinite source of relatively cheap and clean energy, seems to me to be well worth the price.

References

[1]E. Rutherford, Sci. Amer. **225**, 10 (1971).

[2]J. Bartlett, *Familiar Quotations*, 14th ed. (Little, Brown, Boston, 1968).

[3]E. N. da C. Andrade, *Rutherford and the Nature of the Atom* (Doubleday, Garden City, NY, 1964), p. 210.

[4]A. M. Weinberg and R. P. Hammond, in *Proceedings of the Fourth International*

Conference on the Peaceful Uses of Atomic Energy (United Nations, New York, in press); Bull. Atom. Sci. **28**, 5, 43 (1972).

[5] A. M. Weinberg, delivered at "Demographic Policy and Power Plant Siting," Senate Interior and Insular Affairs Committee, Symposium on Energy Policy and National Goals, Washington, DC, 1971.

[6] W. B. Lewis, *Radioactive Waste Management in the Long Term* (DM-123, Atomic Energy of Canada Limited, Chalk River, 1971).

[7] *Disposal of Solid Radioactive Wastes in Bedded Salt Deposits* (National Academy of Sciences–National Research Council, Washington, DC, 1970); *Disposal of Radioactive Wastes on Land* (Publication 519, National Academy of Sciences–National Research Council, Washington, DC, 1957); *Report to the U.S. Atomic Energy Commission* (National Academy of Sciences–National Research Council, Committee on Geologic Aspects of Radioactive Waste Disposal, Washington, DC, May 1966).

[8] Professor Friedrich Schmidt-Bleek of the University of Tennessee pointed out to me that the dikes of Holland require a similar institutional commitment in perpetuity.

[9] L. G. Hauser and R. F. Potter, "The effect of escalation on future electric utility costs" (report issued by Nuclear Fuel Division, Westinghouse Electric Corporation, Pittsburgh, PA, 1971).

Salvaging the Atomic Age

A 1-million-kilowatt, pressurized-water nuclear reactor—the type widely used in the United States today—contains 15 billion curies of radioactivity. This is about equal to the natural radioactivity that accompanies decay of the 4 billion tons of uranium dissolved in all the oceans. Nothing except time can turn this radioactivity off. The radioactivity in a reactor decays only slowly after the reactor is shut down: it contributes about 200,000 kilowatts of heat while the reactor is running, and, depending upon how long the reactor has been running before shutdown, it is still generating 8,500 kilowatts a week later. Unless a large chain reactor is cooled even after the reaction has ceased, the fuel will melt. If it melts, radioactivity will escape from the fuel and may enter the environment.

From the beginning, all of us at Arthur Compton's wartime Metallurgical Laboratory in Chicago, where the first chain reaction was established in 1942, sensed that man had crossed a threshold when he learned how to create radioactivity at will—and on an enormous scale.

Until then, radioactivity was measured in microcuries or millicuries; one gram of radium, costing $50,000, was equal to one curie. (The maximum permissible dose of radium in a human being is one ten-millionth of a gram.) Enrico Fermi, the developer of the chain reactor, and our scientific leader, on occasion would remind us of this. It was not only the Bomb that changed things, he said, it was also the creation of unimaginably large amounts of radioactivity.

The simplest way to ensure that no member of the public was hurt

---

The Wilson Quarterly, pp. 88–112, Summer, 1979. Reprinted by permission.

by the release of radioactivity from a malfunctioning nuclear reactor was to put the reactor in a remote place. To be sure, the first chain reaction, on December 2, 1942, took place on 57th Street and Ellis Avenue in the heart of Chicago's South Side. This tiny venture into large-scale radioactivity gave General Leslie Groves, head of the Manhattan Project, and Arthur Compton plenty of anxiety, but every day counted during the war. To have awaited the completion of the site at Argonne Forest Preserve outside the city would have taken too much time.

But we took it for granted that the large (250,000 kilowatt) plutonium-producing reactors then being planned, as well as a much smaller pilot-plant reactor, would be remotely sited—the latter at Oak Ridge in the hills of eastern Tennessee ("site X"), the former on the huge Hanford Reservation in eastern Washington ("site W"). Most of the other reactors built by the Atomic Energy Commission between 1946 and 1974 were confined to these and several other sites: Savannah River, South Carolina; Idaho Falls, Idaho; and Los Alamos, New Mexico.

All of these places were, at the time, far from population centers; and as the supporting towns, such as Richland, Washington, and Oak Ridge, Tennessee, developed, they were sprinkled with specialists who had daily experience in the handling of large-scale radioactivity and knew how much was dangerous and how much was but a tiny addition to the Earth's all-pervasive natural background radiation.

Had all U.S. power reactors been as secluded as the original Hanford or Idaho Falls or Savannah River reactors, the nuclear enterprise might have avoided many of the problems it has now encountered. We might have had by this time perhaps 25 remote sites, each eventually having as many as 10 or 20 reactors (Hanford at one point had 9 large reactors), each surrounded by a large unpopulated zone, and each ringed at a distance by villages inhabited by people who worked at the plants and for whom the safe handling of radioactivity was a fact of life.

Had the nuclear-energy enterprise remained a government monopoly, as originally prescribed under the Atomic Energy Act of 1946, then the siting and generation of nuclear electricity might have evolved along such lines. Power generation would have been in the hands of the Atomic Energy Commission, or a successor agency, and the electricity so generated would have been distributed by private and public utilities.

But, in retrospect, this could never have happened. The utilities, perhaps stung by the consequences of their foot-dragging on rural electrification in the 1930s, were anxious to forestall the development of a government-operated nuclear version of the New Deal's Tennessee Valley Authority. And this was not a phantom threat: in the early 1950s, Tennessee's Senator Albert Gore was calling (in vain, as it turned out) for federal construction of six large power reactors in the United States. But the Atomic Energy Act of 1954 put nuclear energy into the private sector: Congress allowed utilities to own and operate nuclear power plants, and indeed, encouraged the private sector to design and develop reactors. Today 41 utilities operate some 70 large nuclear power plants.

Producing electricity in remote sites is also awkward. Utility planners build conventional coal- and oil-fired generating plants near cooling water and close to "load centers" (that is, population centers) to reduce the cost of transmission lines. A conventional plant's impact on the environment has, until very recently, been regarded as a secondary matter. As for the possible danger to the public of a plant burning fossil fuels, for a long time this was not an issue, even though the emissions from such plants may cause or exacerbate lung disease. It was all but inevitable, therefore, that nuclear generating plants would by and large be sited as conventional plants had been.

Barriers Within Barriers

Thus, during the early 1960s, Consolidated Edison, which traditionally had put its conventional plants close to New York City, proposed building an underground nuclear plant in the borough of Queens; the proposal was withdrawn only when it became clear that the U.S. Atomic Energy Commission would never license a plant in such a densely populated area.

How could nuclear engineers reconcile the intrinsic danger represented by 15 billion curies in the core of an operating reactor with the necessity of placing the reactor fairly close to population centers?

Several strategies evolved. First, reactors were not allowed *too* close to populated areas. The AEC (and now the Nuclear Regulatory Commission) required "exclusion" zones and close-in "controlled" zones around reactors. No one can live in an exclusion zone, which usually extends about one-half mile from the reactor site; the utility controls access to the controlled zone. Of the 90 or so nuclear sites now in operation or being built, only 13 have more than 25,000 people living within a 5-mile radius, and only 10 have more than 100,000 within a

10-mile radius. To this extent, the original approach to nuclear safety has prevailed, although only a handful of commercial power plants are sited as remotely as Hanford and Savannah River.

But this isolation was clearly not enough. Elaborate engineering devices were developed to place barriers between the environment and those 15 billion curies inside the reactor. The AEC used to speak of three approaches to safety: extremely careful design, to minimize the likelihood of a mishap in the first place; various systems, such as shut-off rods and sensors, to abort a mishap before it gets out of control; and back-up devices, such as the emergency core-cooling system, to cool the reactor and prevent a meltdown, should the regular system fail.

In addition, there are now at least three physical barriers at each reactor between the radioactivity and the world outside: metallic zirconium enveloping the fuel pellets in which the radioactivity is largely generated; the thick steel pressure vessel, along with the pipes that carry the primary cooling water at a pressure of about one ton per square inch; and the now-famous concrete-encased steel dome designed to withstand a pressure of 50 pounds per square inch without leaking. In the event of a core meltdown, should any one of these barriers remain unbreached, little radioactivity would reach the public.

How well had these systems worked before the accident at Three Mile Island, in March, 1979? Pretty much as planned, in American reactors. There were several major accidents, however, as well as many minor incidents:

- In 1961, a serious nuclear excursion, apparently initiated by the deliberate removal of a control rod, killed three operators in a small experimental reactor in Idaho Falls, Idaho; some radioactivity escaped because this reactor, being located so far from people, had no containment shell.
- A loose piece of metal blocked the coolant and caused a partial meltdown at the Fermi fast breeder reactor plant outside Detroit in 1966; both the primary cooling system and the containment shell held.
- At Browns Ferry, Alabama, in 1975, a fire disabled much of the emergency core-cooling system, but enough remained to prevent a core meltdown; no radioactivity leaked to the atmosphere.

Outside the United States, the record, at least in the early days, was not as good. The worst incident was Britain's Windscale fire in 1957.

A plutonium-producing reactor made of graphite caught fire; since the reactor was not surrounded by a containment vessel, some 20,000 curies of radioactive iodine was released, several thousand times as much as was released to the outside in the Three Mile Island accident (10–15 curies).

A properly operating reactor is generally a benign source of energy. It emits no carbon dioxide or sulfur dioxide or particulates. Its radioactive emissions during routine operation are rather less than those from a coal plant of the same output.* The main hazard comes from the 200 tons of uranium that is mined to keep it fueled each year. The mine tailings—material left over after uranium ore is processed—contain about 1,000 curies, but they are usually stored in remote places and much of the radioactivity decays before it is dispersed. Covering the tailings with a foot of earth would reduce even these emissions.

Unlike many critics, I would put disposal of toxic radioactive wastes in the category of lesser problems. This is largely because the high-level, potentially dangerous wastes occupy so little space (two cubic meters per reactor per year) and because, after about 1,000 years, the wastes are no more hazardous than the original uranium from which the wastes were formed. (This uranium is part of nature, and it seems unreasonable to require the sequestered wastes to be less hazardous than the original uranium.) To sequester wastes for 1,000 years simply does not strike me as being beyond reason. After all, cave paintings by Cro-Magnon man have survived for 12,000 years. In Oklo, Gabon, there are ancient underground natural chain reactors that operated for 500,000 years: Many of the fission products and essentially all of the plutonium created in these remarkable phenomena have remained in place, unattended, for almost 2 billion years! Despite the great public concern and political passion generated over nuclear wastes, I view them as a nuisance for which there are many solutions.

Nor do I consider nuclear proliferation, the issue on which so much discussion hinges today, the principal problem of nuclear energy. Most, if not all, of the world's atomic bombs have come from reactors built expressly to make bombs, not from power reactors. Should a country want to make bombs badly enough, it can do so without troubling to build or buy a commercial power reactor. Indeed, I believe nuclear power is rather peripheral to the proliferation issue; our at-

---

*A 1-million-kilowatt coal plant burns 2.5 million tons of coal per year. This coal may have some 10 tons of uranium in it, and this represents about 50 curies of radioactivity in the coal ash and in gases released in the air near the coal plant.

tempts to devise technical fixes for that problem tend to be "allusive and sentimental" (to quote Robert Oppenheimer) rather than "substantive and functional." We can *not* prevent Pakistan or South Africa from making bombs if they are intent on so doing and if their leaders place the manufacture of bombs above other national aims.

The Probability Paradox

The 15 billion curies in an operating reactor, and the possibility of its release, has long struck me as the primary issue, the one on which nuclear energy will stand or fall. Since a serious reactor malfunction is a matter of probability, the issue becomes more and more important as more reactors are built. To illustrate: If the probability of a serious malfunction, in which significant amounts of radioactivity are released, is, say, 1 in 20,000 per reactor per year, then when there are 100 reactors operating, one might expect one such accident every 200 years; but if there are eventually throughout the world 10,000 reactors—as could happen, were nuclear energy to become the world's primary energy source—then, unless the probability of an accident for *each* reactor is reduced, one could expect one such accident every two years.

I do not believe the public in the United States or elsewhere would retain much confidence in an energy system that caused even relatively modest radioactive contamination every two years. Nor does it make much difference where accidents happen: TV converts an accident anywhere into an accident everywhere. For nuclear energy to grow in usefulness, the accident probability *per reactor* will simply have to diminish; the public will have to be prepared to cope with the radiation risk such infrequent accidents might entail; and the media will have to deal with nuclear malfunctions in the same way it deals with other industrial accidents that have comparable impact on health.

Bo Lindstrom, the Swedish aeronautical engineer, pointed out some 20 years ago that air travel faced exactly this dilemma. He argued that, if air travel continued to expand and the accident rate *per passenger mile* held constant, then, by around the year 2000, there would be several serious accidents every day around the world. For the individual passenger, air travel would remain as good a risk in 2000 as it was in 1960. But, he argued, the public's *confidence* in air travel would collapse. Air transport has, of course, become much safer, per passenger mile, than it was at the time Lindstrom made his observations; and,

I suppose, the public and the media have become somewhat inured to occasional air crashes. Both changes were necessary for air transport to survive.

What are the actual probabilities of malfunction in reactors? Before the Three Mile Island accident, all of us in the nuclear enterprise were fairly comfortable with the estimates made by Norman C. Rasmussen of MIT in his famous 1975 study on the probabilities and consequences of a reactor accident.* He estimated that, for a light-water reactor, the probability of a core meltdown that would release at least a few thousand curies of radioactivity was 1 in 20,000 per reactor per year. Most of these incidents would not cause physical damage to the public. A few would, and a very few, estimated at one in a billion reactor years, might be a major catastrophe—3,300 immediate radiation deaths, 45,000 extra cancers, $17 billion in property damage.

Gripping Dramas

Rasmussen himself has set the uncertainty in his estimate of probability at about 10 either way (although the report puts the uncertainty at half this), his estimate of consequences perhaps at 3. That is, the probability of an accident causing significant property damage might be as high as 1 in 2,000 per reactor per year; the probability of the very worst accident 1 in 100 million reactor years. A recent NRC review of the Rasmussen report led by University of California physicist Harold W. Lewis has set even greater uncertainties on the probabilities, although it generally praised Rasmussen's methodology.

Before Three Mile Island, I was comfortable with the record of nuclear energy. The noncommunist world's light-water reactors had amassed 500 reactor years without a meltdown; if one added the U.S. nuclear navy's record, one could roughly triple this—no meltdowns in about 1,500 reactor years. Rasmussen's upper limit of meltdown, about 1 in 2,000 reactor years, was close to being vindicated.

Though I write this before all the returns are in, I believe it is fair to say that Three Mile Island suggests that the probability of accidents that release a few thousand curies may have been underestimated, not so much because of possible engineering deficiencies but because of human error. Closure of two valves, thus incapacitating the auxiliary

---

*U.S. Nuclear Regulatory Commission, Reactor Safety Study: *An Assessment of Risks in U.S. Commercial Nuclear Power Plants*, October 1975. WASH-1400 (NUREG-75/014).

feedwater system, followed by various other malfunctions and operator errors, was not, as far as I can deduce, contemplated in the Rasmussen study.

Yet containment for the most part held. The iodine released was perhaps a dozen curies; the maximum total whole-body exposure to any member of the public was probably less than what we used to accept *every day* as the allowable dose in the early days of the Manhattan project. No member of the public has suffered bodily harm from Three Mile Island.

Chauncey Starr, vice chairman of the Electric Power Research Institute (the research arm of the utilities), has estimated that an incident like Three Mile Island has a fifty-fifty chance of occurring every 400 reactor years. Can nuclear energy survive if such incidents have a 50 percent chance of happening that often? I do not think it can. It is not that people will actually be hurt; it is that people will be scared out of their wits. The drama of the hydrogen bubble in the pressure vessel at Three Mile Island has rarely been matched on TV. And with 200 reactors operating in the United States alone by the 1990s, one might expect a new, gripping TV serial once every few years.

Six Suggestions

Can the nuclear enterprise be redesigned so as to make it acceptable? Can the probability of an incident like Three Mile Island be significantly reduced? And is it likely that the public's (and the media's) reaction to future Three Mile Islands will be more commensurate with the actual damage rather than with their perception of the potential hazard?

In my view, all of this is possible. Indeed, Three Mile Island could be the salvation of nuclear energy. Before the incident, the possibility of a core meltdown was, by and large, the private knowledge of the nuclear and the antinuclear communities. Today, every newspaper reader and TV viewer knows about cooling systems and their malfunction. Best of all, the managers of the electric utilities that operate nuclear plants are now acutely aware that the responsibility entailed in operating a nuclear power plant is far greater than that entailed in operating a fossil-fuel plant.

But "consciousness raising" is not enough. I believe an acceptable nuclear future should have six characteristics: increased physical isolation of reactors, further technical improvements, separation of generation and distribution, professionalization of the nuclear cadre, heightened security, and public education about the hazards of radiation.

Physical isolation

It is unfortunately too late to return to the original siting policy that confined nuclear activities to very remote places. But we can achieve a good deal by confining the enterprise, forever, to those existing nuclear sites that have few people near them. About 80 nuclear sites (operating or being built) currently meet that criterion. Evacuation in an emergency would be relatively easy. More important, everyone living within five miles could be educated about radiation, and each household might be equipped with a radiation detector, much like smoke detectors.

Moreover, the size of the future population within five miles ought to be restricted. This could be accomplished if the area (75 square miles) around each site were properly zoned. If we eventually had 100 isolated nuclear sites in all, this would amount to committing 7,500 square miles to the enterprise in perpetuity. Only a small part of this area, perhaps 200 square miles, would be exclusively reserved for nuclear operations; the rest could be devoted to farming.

Since the number of sites is limited, the generating capacity of each will increase as the nuclear enterprise grows. Eventually, each site may have as many as ten reactors, compared with an average of less than two per site now. Such clustering ought to bring in its wake other improvements. Large sites are likely to have more able people in charge than are small sites. There will develop an organizational memory: Small mishaps on Unit 2 five years ago are not likely to be repeated on Unit 4 today. I speak of this from my experience at Oak Ridge, a large, powerful nuclear center that has always had the logistical strength and organizational memory to contain the damage when accidents have happened.

The sites, like dams, also ought to be invested with an imputation of permanence. If one concedes that the sites are permanent, then one can simply leave the voluminous low-level radioactive wastes (as well as the old reactors) in place until their radioactivity has largely decayed. After 100 years or so, the bulk of the low-level wastes, as well as the old reactors, will be fairly innocuous. Dismantling old reactors after that should be relatively easy. During the decay period, the old reactor buildings might be used to store the other nuclear wastes. In an active, self-contained nuclear complex, maintenance should pose little difficulty.

Technical improvements

As Three Mile Island has shown, the nuclear establishment is still learning. Is it learning fast enough? Will improved back-up systems reduce the probability of failure faster than the number of reactors grows? Surely Three Mile Island will lead to corrections of certain faults in existing pressurized-water reactors. It will also lead to even tougher government regulations. This combination of tougher regulation and improved technology will certainly lessen the likelihood of future Three Mile Islands.

Beyond this is a more fundamental question: Are there types of reactors inherently safer than the common pressurized-water reactor (PWR)? After all, the PWR was conceived as a compact reactor capable of being stuffed into a submarine; its evolution into the mainstay of huge central nuclear power plants on land is still a source of wonder to its original designers. The British, without quite saying so, suggest that their large graphite reactors cooled with gas are less prone to mishaps than is the PWR.*

The Russians continue to build large graphite, water-cooled reactors, as well as reactors like those at Three Mile Island; the Canadians use heavy-water systems. I was a long-time proponent of a completely different reactor type that used fuel that was already in the liquid state, the molten-salt reactor. Is it impossible to return to Square One and try to design a reactor that is more resistant to the so-called China Syndrome? Does the technical community have the nerve, and do the other actors (utilities, government, manufacturers) have the money and the will to design and commercialize a completely new system? Perhaps when the furor over Three Mile Island subsides, we will embark on this uncertain, but possibly rewarding, new path.

---

*Three reactors other than PWRs are used commercially: *gas-cooled graphite*, in which the uranium rods are embedded in a huge block of graphite and are cooled by flowing gas; *water-cooled graphite*, in which water is used as coolant; and *heavy-water*, in which the graphite is replaced by a tank of heavy water—i.e., water in which the ordinary hydrogen is replaced by heavy hydrogen (deuterium).

Generation and distribution

The nuclear system requires a powerful organization if it is to be operated properly. We have in this country about 200 electricity-generating utility companies. Nuclear electricity does not lend itself very well to such fragmentation, nor to small operators. A 1-million-kilowatt power plant often represents a large fraction of the total output of a smaller utility. Twenty-seven of the 41 "nuclear" utilities have but a single reactor. It seems to me that such small nuclear utilities are less likely to maintain the organizational strength and memory necessary to operate a nuclear reactor properly than is an organization that owns and operates many reactors.

If the siting policy that I espouse becomes a reality, it seems natural that the large, clustered sites will be operated by powerful organizations whose main job is operating nuclear facilities. Presumably, most of the sites would serve more than a single utility. Thus, one could envisage the gradual separation of generation from distribution of nuclear electricity, the former being in the hands of powerful organizations—public or private—that *do nothing but generate nuclear electricity*. Such nuclear generating entities would have the technical capacity to supervise every element of the design and construction of their plants. They would be much less at the mercy of the reactor and equipment suppliers than they now are.

It is a delicate, and not very clear, question as to whether the operating consortia should be public or private, whether one would get better (and safer) operation from private organizations policed by the Nuclear Regulatory Commission or from a public Nuclear Energy Authority. There is an inherent tension between *safe as possible* and *cheap as possible*. The conflict manifests itself when a pilot decides to cancel a flight because the weather is bad, even though this costs his company money. Many would argue that a public operator is more likely to weigh his decisions on the side of safety. But public authorities seem to me to be harder to regulate than private ones.

After Three Mile Island, every nuclear utility, not to mention reactor manufacturer, must realize that its very existence may depend on avoiding incidents of this sort. This must be powerful medicine for clearing one's brain of a possible confusion as to which comes first, safety or continuity of electricity supply.

Professionalization of the nuclear cadre

The pilot of a transatlantic Boeing 747 is paid about $100,000 per year, perhaps 50 percent of what the president of his airline gets. The superintendent of a nuclear plant gets $40,000 per year, about 20 percent of what his president gets. The pilot and the operator bear a heavier burden of direct responsibility for people's lives than do their respective bosses. In the case of the pilots, this is more or less acknowledged in their pay; in the case of a plant superintendent, it is not. Why?

I believe the answer is to be found again in the mistaken belief among utility executives that a nuclear plant was just another generating station. The pay scales—indeed, the whole conception of training and expertise—tended to be strongly influenced by this perception. Moreover, a utility manager found it awkward to pay operators of one kind of plant much more than he paid operators of another.

But the responsibility borne by the nuclear operator is so great that he and his staff must be regarded—and trained—as an elite. They must constitute a cadre with tradition, competence, and confidence. Is it possible to get people of such quality for jobs that are essentially very boring? This is the same problem faced, say, by the pilots on the Eastern Airline shuttle between Washington and New York or by the anesthesiologist during a routine operation. We deal with this ennui with money, with status in the community, with shorter work schedules. This is the very least we can do with nuclear operators.

That is why I have so strongly urged cluster siting and separate generating entities: Both would be more conducive to creation of the professionalized corps necessary to keep the nuclear enterprise out of trouble. At a nuclear center, there will be *many* people to choose from when vacancies arise. There will be a general ambience of expertise. And an independent generating entity can pay its employees salaries that are not bound by the locked-in traditions of coal-fired utilities.

Heavy security

The nuclear enterprise will always demand far greater security than the fossil-fueled enterprise. Terrorists and saboteurs can merely incapacitate a fossil-fueled plant; in nuclear plants, they can, albeit with some difficulty, produce serious accidents. This is another reason why cluster siting is important. It is easier to guard 10 reactors on a single site than 10 reactors on separate sites.

Public education about radiation

None of the above measures will ensure the survival of nuclear energy unless the public and the media come to accept the risk of radiation as no different from the risk of other noxious substances that are products of our technology, particularly agents such as mercury or polychlorinated biphenyls that persist and sometimes (as at the chemical plant explosion in Seveso, Italy) interdict land.

Why, after all, did Three Mile Island create such extreme concern, especially since not one member of the public has been harmed by it or, for that matter, by the operation of any other U.S. commercial nuclear reactor? Why was Three Mile Island the biggest story of the year when the collapse of the Grand Teton Dam in 1976, or even the collision of the jumbo jets in Tenerife in 1977, vanished from the front pages in a few days?

I see several reasons for this seeming double standard. The potential for a disaster was there, although exactly how close we came will have to await the outcome of the current investigations. Since its dimensions could not be gauged and the whole situation was so completely novel, the crisis provided the classic ingredients of high media drama. The fear of possible radiation-induced death goes deep. Radiation *is* mysterious: It cannot be sensed, you can't see it, yet it can kill you.

The estimate of the hazard of radiation is clouded by bitter scientific dispute. In particular, there is the strongest kind of disagreement among scientists as to the effect of very low levels of radiation, even levels as low as our natural radiation background. Most of the estimated delayed cancer deaths associated with so-called hypothetical accidents are supposed to be caused by exposures well below the occupational limits. If one assumes that any extra radiation, however small, causes cancer, then, if millions of people are exposed, some extra cancers will result. But if, as I believe, low-level radiation is nowhere near as dangerous as, for example, television newsmen seem to think, then the public's (and, perhaps more important, the media's) reaction to the possibility of such irradiation may be far more restrained.*

The whole question of low-level radiation is so critical to public acceptance of nuclear energy that I consider this a leading, if not *the* leading, scientific issue underlying the nuclear controversy. Unfortunately, since the effects (if any) are so rarely seen because the exposures are so small, the issue may be beyond the ability of science to decipher. Fortunately, we do have a standard—natural background radiation—with which to compare additional exposures. At Three

---

*The controversy over low-level radiation is examined in *The Effects on Populations of Exposure to Low Levels of Ionizing Radiations*, the report of the Committee on the Biological Effects of Ionizing Radiations, National Academy of Sciences (1979). The committee divided sharply on the issue; a dissenting report was appended.

Mile Island, the total dose to the population was about 1 percent of natural background—a level at which no effects can be seen.

Unless changes are made that restore the public confidence, the Nuclear Age will come to a halt as the present reactors run their course, and we shall have to revert to the energy strategies that were available before fission was discovered. What are the alternatives to fission? Aside from conservation, which has its limits, there are only four: geothermal, fusion, fossil, and the various forms of solar energy.

The potential of geothermal energy—from natural steam or hot dry rocks—is relatively small; if we are to contemplate a world that has many more people, and that uses, say, three times as much energy as we now use, geothermal can hardly help.

As for fusion, despite the optimism that prevails among scientists working in the field, it seems to me that the possibility still remains just that—a possibility. The fuel, deuterium and tritium (isotopes of hydrogen), is all but inexhaustible, yet the engineering remains formidable. Moreover, fusion is not devoid of radioactivity. To be sure, there is 100 times less in a fusion reactor than in a fission reactor. But, as Three Mile Island suggests, if fusion is to be acceptable, it too will require a public that understands the relative hazards of radioactivity. Thus, my view about fusion is agnostic—let's work on it, but let's not count on it.

Fossil fuel is, of course, what we shall turn to in increasing amounts whether or not we have fission. But if we had a moratorium on new fission plants beginning in 1985, we might, in the United States, have to burn about a billion tons more of coal by the year 2000 than if we had no such moratorium. And as for oil, the political pressures might become quite intolerable, should our need for the world's oil increase drastically.

Nor is fossil fuel a benign source of energy. Even a *properly operating* coal plant emits noxious fumes. The dangers from burning fossil fuels are undramatic—deaths from coal-mine accidents, black lung among miners, bronchial troubles downwind of coal-burning plants. By contrast, the dangers of a nuclear plant are localized and dramatic, even though, as Three Mile Island has shown, a nuclear plant can suffer an extraordinary amount of damage without anyone being hurt. Fossil fuels also pose a large-scale worldwide threat comparable to that of proliferation of nuclear bombs. I refer to the accumulation of carbon dioxide (CO_2) in the atmosphere. Most climatologists (though not all) believe that doubling the CO_2 may increase the average surface temperature of the Earth by about 2 °C; this would diminish the equatorial-polar temperature gradient, which drives the wind system, by about 10 °C.

Not all the ensuing changes need be bad. But the doubling of CO_2 in the atmosphere, which could happen by, say, the year 2050, represents an unprecedented climatological experiment. It might cause the seas to rise, turn deserts into grasslands, grasslands into deserts. If this is a real possibility, then would continued burning of fossil fuel be a responsible course, even if fossil fuels were inexhaustible (which they aren't)?

Which leaves us with solar energy, including hydro, wind, waves, biomass, and ocean thermal gradients as well as direct solar. Solar energy is immense, environmentally benign, the darling of the people (no one is against solar energy). But it is also intermittent and, insofar as one can tell, expensive. If what we contemplate is an all-solar world, not one in which small household solar water heaters are backed up by electricity from the local utility, then we must come to terms with the intermittency of solar energy. Either we adjust our lives to a Sun that does not always shine, or else we arrange for storage—perhaps with auxiliary engines operated on alcohol, or electric batteries, or perhaps by hydrogen generated photoelectrically. Overcoming intermittency seems to be very expensive, though how expensive I cannot say. What seems clear is that an all-solar society is almost surely a low-energy society, and one in which energy will be a good deal costlier than it is now.

The Faustian Bargain

To my mind, the only alternative (or perhaps adjunct) to a solar society is the one based on fission—at least if one concedes that fossil fuels are limited, or that the CO_2 danger must be taken seriously, or that fusion will forever evade us. But such a fission future might involve several thousand U.S. reactors, and one must then come to terms with the problem I alluded to earlier: Even though the probability per reactor of a serious accident is small, the frequency of accidents may become too high for the public to tolerate when the system becomes large.

Thus, if a solar society can be made to work, by all means let us work hard to achieve it. I favor pushing solar technology as hard as we can. But let us not mislead ourselves. Solar cannot take over very much of the load for a long time, if ever; and a solar society will not be the utopia many advocates perceive it to be, even if some very major improvements in energy storage and photoelectric conversion are achieved.

But suppose we do not achieve these technological breakthroughs: Can we put a price on solar energy at which we would prefer it over nuclear because nuclear is handicapped by its radioactivity? There are many who would abolish nuclear in favor of solar *whatever* the cost. This I cannot view as a rational response. But neither can I say how much extra one should pay for solar to avoid the disadvantages of nuclear. And it is not impossible that the price one must pay for an acceptable nuclear system—with its better technology, higher-paid personnel, and tighter security—conceivably could price nuclear out of the market.

About eight years have passed since I first referred to nuclear energy as a Faustian Bargain. I have since been corrected both by nuclear advocates (who prefer Prometheus to Faust) and by scholars (who say that Goethe's Faust didn't really make a bargain at all). Nevertheless, what I meant was clear: nuclear energy, that miraculous and quite unsuspected source of energy, demands an unprecedented degree of expertise, attention to detail, and social stability. In return, mankind has, in the breeder reactor, an inexhaustible energy source.

Three Mile Island has undoubtedly turned many away from nuclear energy, has reinforced their belief that nuclear energy is simply too hazardous. Three Mile Island for me has a rather different significance. I have often said that Goethe's Faust was redeemed—"Who e'er aspiring, struggles on,/For him there is salvation"—and that mankind, in its striving, will finally master this complex and unforgiving technology.

My antinuclear colleagues retort that this is foolish technological optimism—that mankind is imperfect, and that anything that can go wrong will go wrong. But mankind is also ingenious, and the history of the two worst American reactor accidents—Browns Ferry and Three Mile Island—demonstrates this. In both cases, the accident was precipitated, or at least exacerbated, by human error: a lighted candle at Browns Ferry, closed valves at Three Mile Island. In both cases, the operators used their ingenuity to contain a nasty situation. And in *neither* case was anyone harmed by excess radioactivity. It is their cynical denial of human ingenuity and uncompromising acceptance of human fallibility that is the main weakness of the more strident nuclear opponents.

I am not an uncritical advocate of nuclear energy. I believe the enterprise needs fixing if it is to survive. Nor can the nuclear enterprise wait too long before its managers demonstrate to the public that they recognize this fact. I hope those of us who believe that nuclear energy cannot be lightly cast aside will do more than simply restate our faith in the technology. We must come up with positive and convincing initiatives that prove to the public that the lessons of Three Mile Island have been learned. To do less will commit us to energy options whose inherent difficulties, though not as dramatic as those of nuclear energy, could in the long run be even more serious.

Burning the Rocks Forty Years Later

I

Phil Morrison could hardly contain his excitement as he showed me his calculations: If uranium were burned in a breeder, the energy released through fission exceeded the energy required to extract the residual 4 ppm of uranium from the granitic rocks. The breeder therefore represented an essentially inexhaustible source of energy! H. G. Wells' 1914 dream of "a world set free"[1] through inexhaustible nuclear energy was, in principle, attainable.

The time was 1943, the place the wartime Metallurgical Laboratory in Chicago. Harrison Brown was then working in the Laboratory's Chemistry Division on methods for separating plutonium from irradiated fuel; he soon moved to the Clinton Laboratories in Oak Ridge to serve as Assistant Director of the Chemistry Division there. He, as well as the other chemists at Clinton, were aware of the ultimate potential of the breeder; and, after Brown moved to Caltech, he and L. T. Silver studied the chemistry of extracting residual uranium and thorium from granite.

Brown's study of the use of low-grade uranium ores was motivated by his speculations about the shape of mankind's future. In *The Next Hundred Years*,[2] which he coauthored with James Bonner and John Weir, and which was published in 1957, one finds an asymptotic en-

Written in 1983 and published in *Earth and the Human Future, Harrison Brown Festschrift*, Westview Press, Boulder, CO, 1986. I am very grateful to my colleagues, H. G. MacPherson, A. M. Perry, I. Spiewak, and R. Rainey, for critically reviewing my manuscript.

TABLE 1. Projected energy input pattern for the year 2060, assuming a world population of 7×10^9 (after Brown, Bonner, and Weir).

Source	Energy input (TWyr/yr)
Solar energy (for 2/3 of space heating)	14
Hydroelectricity	3.8
Wood for lumber and paper	2.4
Wood for conversion to liquid fuels and chemicals	2.1
Liquid fuels and petrochemicals produced via nuclear energy	9
Nuclear electricity	32
Total	63.3

ergy budget for a world of 2060 with 7×10^9 souls. I have reproduced this budget in Table 1.

Brown's total of 63 terawatt years per year (TWyr/yr) would now be regarded as too high; at least the 1981 IIASA Report, *Energy in a Finite World*,[3] suggests a total energy budget (for 2030 rather than 2060) that is less than half of Brown's 63 TWyr/yr. Noteworthy was Brown's notion that two-thirds of the primary energy would be provided by nuclear reactors. If each reactor produced, say, 2,500 megawatts of heat and 1,000 megawatts of electricity, Brown's world would require some 17,000 such reactors—about 35 times as many as are now in operation or under construction. Thus, for him, a very large source of uranium, such as was to be found in common granite, was a key to the long-term well-being of the human race.

Brown and Silver reported their results in 1955 at the first Geneva Conference on Peaceful Uses of Atomic Energy.[4] Their main finding was that, of the 4 ppm of uranium and 10 ppm of thorium in granite, about 25 percent was surprisingly easy to leach with dilute acid. Thus, the energy balance for extracting uranium from granite almost surely was favorable, if the uranium were burned in a breeder in which 50 percent of the ^{238}U underwent fission. Moreover, uranium from common rock, even at Brown and Silver's estimated price of $550 to $1,200 per kilogram (in 1955 dollars!) might be affordable if it were burned in a breeder. To quote Brown and Silver: "There is ample uranium and thorium in the igneous rocks of the earth's crust to power a highly industrialized world economy for a very long period of time."

How "inexhaustible" was this source of energy? In the crust, down to a depth of 1 mile, there is 2×10^{12} tons of uranium; in the seas, another 4×10^{9} tons. Brown's 17,000 breeders, each burning about 1 ton of uranium per year, could be supplied with uranium from rocks for millions of years. Even if only the sea were mined, there would be fuel to last for 2×10^{5} years. Moreover, as Bernard L. Cohen has since pointed out,[5] the uranium leaching naturally from the rocks replenishes the uranium in the sea at the rate of 30,000 tons per year. Even without grinding granitic rocks, we might have an inexhaustible energy source in the breeder reactor, provided the uranium from the sea could be won at a reasonable price.

II

That the breeder was an inexhaustible energy source really was not fully and officially recognized until much later. Even today, one encounters the partially correct assertion that the breeder increases the potential amount of energy from uranium about 50-fold to 100-fold, this being the ratio of practically minable ^{238}U to fissionable ^{235}U. This assessment ignores the enormously greater amount of low-grade uranium and thorium ore that can be burned in a breeder but not in a conventional burner reactor. Only with the creation of the Energy Research and Development Administration was the breeder (along with fusion and solar energy) officially designated as an "inexhaustible." Yet implicitly, ever since Walter Zinn produced electricity from fission for the first time in the Experimental Breeder Reactor (EBR-I) in 1951, pursuit of the breeder had been firm national policy. Fusion had been accepted as a national goal at about that time; and even solar energy received a little support from the National Science Foundation as early as 1953.[6] Thus the United States, as well as much of the rest of the world, has been committed to the goal of inexhaustible energy, via fusion, fission, and, to a lesser degree, solar, for 30 years. We now spend more than a billion dollars per year in the United States alone on the three inexhaustibles. About $600 million per year goes for direct support of breeders, and a like amount goes for fusion; about the same amount, mainly in the form of tax rebates, supports various forms of solar energy.

When would an inexhaustible energy source—particularly the fission breeder—be needed? None of us could give a totally convincing answer 30 years ago, nor can we give a totally convincing answer today. The timing of the breeder depended both on how much cheap uranium would be found and on the demand for electricity. When Walter Zinn demonstrated the first breeder prototype, the known reserve of uranium (at $20 per kilogram of uranium, in 1952 dollars) in the United States amounted to 7,300 tons. Today, OECD estimates the reserve in the non-Communist world to be 5×10^6 tons of uranium at less than $130 per kilogram (that is, $40 in 1952 dollars); of this, 1.7×10^6 tons is in the United States. A 1,000 megawatt light-water reactor (LWR), operating at 70 percent capacity, uses 4,300 tons of uranium during its expected 30-year lifetime. The 5×10^6 tons of uranium therefore would support about 1,000 such reactors for 30 years. Were the world's uranium supply 20 times as large, fission in LWRs could still support an energy system as large as Brown visualized for only 30 years. No wonder that Eugene Wigner,[7] the godfather of the breeder reactor, argued that, without the breeder, nuclear energy was not all that important a source of energy: that nuclear energy would eventually die out unless breeders (or their equivalent) were developed. He was careful to keep the technical paths to breeders open. As he put it in the late 1950s, it was too early to say whether breeders would gradually evolve from nonbreeders or would arise as a completely independent technology. Moreover, we could not choose then between slow neutron breeders based on the thorium-^{233}U cycle and fast neutron breeders based on the uranium-^{239}Pu cycle.

Though most countries took for granted the necessity for breeder development, Canada chose a different course. Since Canada's requirement for nuclear electricity was relatively small and its reserve of uranium was enormous, conservation of uranium did not seem as important as simplicity of the fuel cycle. *Breeders Are Not Necessary!* was the title of a famous paper published by W. Bennett Lewis in 1963.[8] Lewis's argument was simple: It is not necessary to burn 50 percent of all the ^{238}U to utilize a significant fraction of the low-grade ores. A system that burned only 3 to 15 percent of the uranium and thorium—still 5 to 25 times as much as was burned in an ordinary light-water reactor without recycling the unburned ^{235}U—would suffice to burn all but the very lowest grade ores. And Lewis's heavy-water reactor fueled with ^{233}U, thorium, and uranium could achieve this lesser aim with a reactor technology not very different from the well-established technology of the Canadian Deuterium Uranium (CANDU) reactor, though very frequent recycling of the unburned uranium fuel was required to achieve such performance. This historical debate still rages between those who cannot conceive of a long-term nuclear energy without some kind of breeder and those who argue that a 100- or 200-year solution to the nuclear energy problem—as provided by improved CANDU or even LWR reactors with recycling—was far more realistic than was a full-fledged breeder, such as the liquid-metal fast breeder reactor (LMFBR) or the molten-salt breeder reactor (MSBR).

One's attitude toward breeder development depends on which problem one seeks to solve: the relatively short-range one of providing a source of electricity that is competitive with other sources of electricity, or the very long-range one of providing an inexhaustible energy source. In the first instance, cost was of the essence; in the second, though cost was not unimportant, it was not necessarily the most important factor. As a practical matter, these two goals have had to be confounded. Though the goal of inexhaustible energy is as lofty as any technological goal, one could not give a convincing reason for pursuing this goal on a short schedule, unless the breeder—or the breeder in combination with nonbreeding burner reactors—could produce competitive power. Twenty years ago, the breeders appeared capable of producing economic power. At that time, the Oyster Creek reactor was built for about $120 per kilowatt of electricity. Were a breeder to cost 50 percent more than Oyster Creek—that is, $180 per kilowatt of electricity—this difference could be made up if the fuel-cycle cost for the breeder were a mere 2 mills per kWh cheaper than the fuel-cycle cost for the burner. The price of raw uranium used in an LWR would have to increase by $60 per kilogram for the fuel cycle in an LWR to rise an additional 2 mills per kilowatt-hour: the "compensation price of uranium"—that is, the rise in the price of uranium that would compensate for the difference in capital cost between breeders and LWRs—would be $60 per kilogram higher than the current price of about $60 per kilogram. At the time, breeders at $180 per kilowatt seemed like an entirely plausible goal; and the American response to Canada's claim that breeders are not necessary was "But electricity from a breeder should be cheaper than from an LWR. Breeders would be introduced naturally because they were a cheap way to make electricity; that they also provided a path to inexhaustible energy is secondary. In any event, power from breeders, unlike power from LWRs, is insensitive to the cost of uranium."

The escalation in capital costs of reactors has changed this calculus. Today, a large LWR costs, on average, about $2,000 per kilowatt of electricity. Were a LMFBR to cost 50 percent more than a LWR—that is, $3,000 per kilowatt of electricity—the compensation price of uranium would come to about $900 per kilogram, a price we generally consider to be much higher than the cost of uranium from the vast deposits in the Chattanooga shale. Thus, unless the capital cost of the breeder can be reduced—to perhaps 1.2 times the capital cost of a LWR—energy from breeders simply would not compete with energy from nonbreeding LWRs for a very long time. The original argument for *quick* deployment of the breeder has therefore been compromised by economic realities.

III

These arguments were raging in 1962 when the Atomic Energy Commission published its *Report to the President*,[9] which set forth the nuclear energy strategy of the United States. Central to the strategy was the assumption that electricity would continue to grow into the next century at its historical rate of 7 percent per year. By the year 2000, almost 40 years from the time the report was issued, 734 gigawatts of electricity (GWe) were expected to be nuclear, according to the 1967 sequel to the 1962 report. Were this projected demand to be supplied entirely by standard LWRs, by the turn of the century the U.S. domestic demand for uranium would have amounted to about 100,000 tons of uranium per year. Since the U.S. reserve of low-cost uranium was around 2×10^6 tons, it was clear that reactors that used uranium more efficiently—either advanced converters or breeders—were necessary if nuclear energy were to survive for even 50 years without a steep rise in the price of fuel.

The *Report to the President* recommended that both the uranium and the thorium breeder cycles be pursued—the first by means of fast neutron breeders, the second by means of slow neutron breeders. Although a slow neutron breeder, the light-water breeder reactor (LWBR), was constructed and operated by Admiral Rickover, the preponderant technical sentiment throughout the world favored the LMFBR. This reactor, cooled by sodium, represented a technological thread quite distinct from that of the mainline LWR. Thus, even by the middle of the 1960s, it appeared that Wigner's uncertainty about whether breeders would develop as outgrowths of nonbreeders or as completely different devices seemed to be settling in favor of the latter possibility.

TABLE 2. Liquid-metal fast breeder reactors.

Country	Reactor	Power (MWe)	First date of operation
France	Phenix	233	1973
	Super-Phenix	1,200	1984
Germany	SNR-300	295	1986*
Japan	Monju	300	1987
United Kingdom	Dounreay	250	1976
USA	EBR-II	15	1958
	FFTF	440 (thermal)	1980
	Clinch River	330	1990*
USSR	BN-350	350	1973
	BN-600	600	1980

*Canceled.

Skeptics remain within the nuclear community who continue to espouse breeders other than the LMFBR. The group at Oak Ridge (to some degree influenced by Harry Brown's interest in reactors fueled with uranium in fluid form), designed and built four small reactors in which a liquid fuel was used. In the two earlier reactors, uranyl sulfate was dissolved in water; in the later ones, a molten uranium fluoride salt (that could also bear the fertile thorium) was used. In such fluid-fuel reactors, the chemical reprocessing of bred ^{233}U is relatively simple, since the bred fuel is already in the liquid state. The developers of the high-temperature gas reactor (HTGR) have proposed a gas-cooled fast breeder. And, as the LWBR attests, ideas for breeders based on the technology of light-water reactors continue to command attention.

The main line of breeder development, however, is the sodium-cooled LMFBR. Altogether, about a dozen such reactors have been built or are under construction. The current breeders (as of 1983) are listed in Table 2. Of these, the most successful, at least in the West, has been the French Phenix. It has closed its fuel cycle, and has demonstrated a substantial breeding gain.* Moreover, the fuel elements can sustain almost 10 percent burn-up of the uranium–plutonium oxide

*Breeding gain is the number of atoms of plutonium produced in the breeder for each atom of plutonium burned, minus one. Thus, in Phenix, about 1.13 atoms of plutonium were recovered per atom burned; the breeding gain is therefore 0.13.

mixture contained in them rather than the two or three percent regarded as the limit in earlier breeders. This means that, on average, a plutonium atom need be recycled only two or three times before it is fissioned. The chemical recycling can therefore be more conveniently decoupled from the breeder than was believed to be possible in the early breeders such as EBR-I, which required 10 rather than two or three recycles per fissioned atom of plutonium. Thus, the recycling for Phenix is little more demanding than that for ordinary LWRs.

IV

Just as the technology of the breeder is being demonstrated, many powerful voices urge the dismantling of the enterprise. The arguments against the breeder are mostly extensions of the arguments against nuclear power: reactor safety, waste disposal, cost, and proliferation. One is hard put to understand why the breeder is so often singled out for particular opprobrium by those who dislike nuclear energy. After all, the breeder is hardly less safe than a LWR; waste disposal is equally easy or difficult for breeder or nonbreeder; and the danger of proliferation from breeders is neither greater nor smaller than that of proliferation from nonbreeders with recycling—that is, advanced converters, which also require recycling.* If one assumes, as do I, that LMFBRs are adequately safe, that wastes can be sequestered satisfactorily, and that proliferation is not uniquely a problem for breeders, then I can see little merit in the noisy outbursts against breeder technology. Indeed, I suspect that the attack on breeders is really an attack on nuclear energy since, unless breeders or their equivalent are developed or we learn how to extract uranium from the sea cheaply, nuclear energy is eventually self-limiting.

As for capital costs, an LMFBR can no longer afford a capital cost 50 percent higher than the capital cost of today's LWRs. Though the nuclear community expects to reduce the cost of future breeders, this remains an expectation, not a reality. Unless breeders become cheap enough, burner reactors, not breeders, will be built until the compensation price of uranium is reached. Should this price be as high as, say, $500 per kilogram, the time for the breeder could be fairly far in the future.

---

*Weapons-grade plutonium is easier to milk from the breeder; however, I regard this as secondary, since bombs can be made from reactor-grade plutonium also. Note added in 1991: Iraq's path to a bomb was based on isotope separation, not on breeders!

The proliferation issue deserves more discussion. The objection to the breeder as being particularly vulnerable to proliferation is based on the breeder's requirement for a reprocessing plant. Thus, so the argument goes, a country, by pursuing a breeder with its chemical recycling of unburned plutonium, could acquire a plutonium capability entirely legitimately. Should it decide later to divert material for bombs, it could do so without having to go to the trouble of building a clandestine reprocessing plant.

This argument has merit, in principle. However, I do not believe it takes full account of the improvements in breeder fuels that tend to weaken the coupling between breeder and chemical recycling plant and that open new possibilities for controlling the flow of plutonium. As long as breeders required much more recycling than did LWRs with recycling, the breeder–recycle complex could be viewed as affording greater opportunities for diversion of plutonium to would-be proliferators than did LWRs. But with recycling in the breeder now being no more frequent, per megawatt year of energy production, than recycling in the LWR, centralized, internationally supervised recycling centers can be regarded as practical. A 5,000-ton-per-year chemical plant could service about 150 breeders, each of 1,000 megawatts electric (MWe) capacity. A world of, say, 3,000 breeders would require about 20 such plants. These plants would bear the same relation to the breeder as isotope-separation plants do to today's LWRs. The bred fuel returned to breeders from these plants could be spiked with radioactivity; an additional radiochemical cleansing would be required before the material could be used in bombs.

Of course, all this presupposes that the recycling is conducted in heavily supervised centers. Yet 20 recycling plants throughout the world hardly strike me as being impossible to supervise. Moreover, the fuel-cycle costs for plants of this size are much lower than they would be for smaller plants. There would therefore be powerful economic incentives for a country that builds a few breeders to send its spent fuel to such a center for reprocessing, rather than building its own much smaller chemical reprocessing plant. Economics, therefore, favors a breeder regime that is proliferation-resistant. Nevertheless, one must concede that any reprocessing plant, whether for breeder or LWR, is susceptible to misuse; prevention of such misuse would require international agreements and inspection.

Can the economic incentives to reprocess centrally be made so strong and so evident that any attempt to build one's own reprocessing plant would, *prima facie*, be regarded with suspicion? Probably not, when only a few breeders are deployed, since the economy of scale for the needed reprocessing plant is then nonexistent. On the other hand, breeders will hardly be deployed at first except in the most advanced countries—the weapons states plus Germany and Japan; and for them, the whole issue of proliferation is clearly a political one, not a technical one. And when many breeders have been deployed, economy of scale should be so clear that a small "legitimate" reprocessing plant simply could not be justified except on grounds other than economics—for example, a desire for autarky, or broad considerations of national security. And, I suppose, additional incentives to reprocess fuel centrally can be imagined. Perhaps the most powerful would be an offer to handle wastes at the reprocessing site rather than returning the wastes to their country of origin, though the difficulties of such a proposal are manifest. The Soviet Union has imposed such a quid pro quo in handling spent fuel from Soviet-built reactors in foreign countries.

V

In contemplating the breeder's future, we must address two possible contingencies: either we will continue to depend on a large contribution from nuclear energy (in which case we must assure the supply of nuclear fuel, either from breeders or from some other source) or we will turn away from fission and depend eventually on other inexhaustible energy sources.

As for the second contingency, the only inexhaustibles other than the breeder are solar and fusion. Solar continues to be expensive; and fusion's feasibility, both technically or economically, is anything but assured. Under the circumstances, rejection of the breeder, on the grounds that these other options are known to be available, seems to me to be unjustified, even foolish.

On the other hand, at least four paths to almost inexhaustible nuclear energy must be considered more seriously today than they had been at the time the breeder was first proposed. These four are improved converter reactors; exploitation of low-grade ores, particularly sea-water; fission–fusion hybrids; and accelerator breeders. I shall discuss each of these briefly.

Improved converters

A 1,000-MWe LWR operating at 70 percent capacity uses 4,300 tons of uranium over its 30-year nominal lifetime. Improvements in utilization of fuel in LWRs range from 15 to 30 percent with relatively small rearrangements of the lattice to factors of almost 2 with full recycling and recovery of unused fuel at the end of the cycle. Thus, reactors that use about 100 or even 75 tons of uranium per year rather than the 150 tons per year now used may well be in the cards. Each such improvement puts off the time when the supply of uranium finally gives out. Such improvements are inevitable, regardless of whether breeders are developed or not; and improved converters will surely pose formidable economic competition for breeders, at least during the next 50 years or so. On the other hand, as the conversion ratio* increases in a converter, so does the frequency of recycling. Whether Bennett Lewis's goal of a reactor that uses, say, 50 tons or less of thorium and uranium per year can be realized at a competitive price remains rather uncertain. And, of course, the advanced converter with recycling poses the same threat of proliferation as the breeder.

Low-grade ores

The usefulness of very low-grade ores in converter reactors depends not only upon the cost of extracting uranium but also upon the net energy that can be derived from these sources. Both Morrison's and Brown and Silver's original calculations of the energy required to extract uranium from granites were crude; since then, there have been many studies of the net energy derived from burning uranium derived from low-grade ores. Perhaps the most exhaustive study was made in 1975 by R. M. Rotty, A. M. Perry, and D. B. Reister at the Institute for Energy Analysis.[10] These authors found that the net energy ratio, that is, the ratio of energy produced to energy required to extract and enrich the uranium and to construct the reactor, was comfortably greater than 1, even for a standard pressurized water reactor (PWR) fueled with uranium from Chattanooga shale. A few results are quoted in Table 3.

At very low ore grades, the net-energy ratio ought to be inversely proportional to ore grade, since the energy required to extract the uranium becomes very large. Thus, if the net-energy ratio is 3.4 for an ore containing 60 ppm, one might estimate that the net-energy ratio for a standard PWR becomes 1 at about 20 ppm; for a PWR with recycling, it becomes 1 at around 12 ppm. These numbers are rough estimates, of course. Perhaps most significant is that uranium, even

*Conversion ratio is the number of new fissile atoms produced per fissile atom burned. In a converter reactor, this ratio is less than one; in a breeder, it is greater than one.

TABLE 3. Lifetime net-energy ratio for PWR, 1000-MWe plant, operating at 75% capacity.

Ore grade (ppm)	Pu recycle	Net energy ratio
1760 (conventional ore)	No	9.6
60 (Chattanooga shale)	No	3.4
60 (Chattanooga shale)	Yes	4.9

from Chattanooga shale at 60 ppm, can be used with a comfortable net-energy ratio in conventional reactors. The Chattanooga shales contain 6 million tons of uranium that is claimed to be available at perhaps \$500 per kilogram; and uranium at this price would add little more than 1.5¢ per kilowatt-hour to the cost of power. If one takes these claims at face value, one would conclude that, for the next 100 years, the United States could provide much of its electricity from more or less conventional reactors, at least if the demand remains less than, say, 1,000 GWe, and if we are content with allowing our electricity to rise in price by less than 2¢ per kilowatt-hour. All of this is based on the assumption that uranium can be extracted from the Chattanooga shale at around \$500 per kilogram and that the enormous despoliation of Appalachia caused by such extensive hard-rock mining would be acceptable. To provide 100,000 tons of uranium per year from the Chattanooga Shale, about 4 billion tons of shale (assuming 70 percent recovery) would have to be mined. This is more than 5 times the total amount of coal mined throughout the United States!

This environmental catastrophe could be avoided if breeders were used or if uranium could be extracted from sea water economically. Though recent studies summarized by OECD claim that uranium from the sea could be recovered for \$500 to \$1,000 per kilogram, such estimates must be regarded as speculative. Nevertheless, were we certain that uranium from the sea would cost no more than \$500 per kilogram, the case for the breeder would be much weakened.

Fission–fusion hybrids and the accelerator breeder

Traditionally, the breeder has been looked upon primarily as a power producer that incidentally produced a relatively small amount of fissile material as a by-product. Modern embodiments of LMFBRs are expected to double their 5-ton inventory of plutonium every 15 years or so—enough to keep up with an expansion of the nuclear power system at about 5 percent per year. Alternatively, in a steady-state system that included both breeders and LWR converters, the bred plutonium from the breeder could be fed back into the LWR; a 1,000-MWe breeder could serve as a "cow" for one conventional 1,000-MWe LWR, and perhaps for three improved converters.

In principal, production of fissile material by much more efficient "cows," based either on fusion or on accelerators, is conceivable. The so-called fission–fusion hybrid would utilize the neutrons produced as a by-product of the deuterium–tritium fusion reaction to convert ^{238}U into ^{239}Pu or ^{232}Th into ^{233}U. The accelerator breeder would utilize the neutrons produced by bombardment of a target with high-energy protons to generate fissile material. One fission–fusion hybrid or accelerator breeder could service perhaps 15 LWRs whereas only one LWR could be serviced by a conventional breeder. Thus, even if the cow cost twice what an LWR cost, the capital cost of a system consisting of one cow plus 15 reactors would be only 13 percent higher than that of an LWR. The cow would bear the same relation to the LWRs as diffusion plants now do; it would be not so much a power producer as a fissile-material factory.

Cows for the production of fissile material, whether based on fusion or on accelerators, face formidable problems. To be sure, the electronuclear breeder is feasible in principle, but its cost is very difficult to ascertain. As for the fission–fusion hybrid, there remain questions about its scientific feasibility, let alone its practicality. Nevertheless, these possibilities cannot be dismissed out of hand; and during the coming decade,* we shall be able to say much more clearly than we can now how practical the fission–fusion hybrid or the electronuclear breeder will be.

VI

Events seem to have caught up with the breeder: a combination of lessened demand, higher capital cost, and additional sources of uranium, as well as the prospect of alternative inexhaustible sources of fissile material from fission–fusion hybrids or accelerator breeders, would seem, at the very least, to push the time for deployment well

*Note added in 1991: We still cannot say how practical the fission-fusion hybrid or the electronuclear breeder would be.

into the next century. With solar energy and fusion being seriously developed, fission energy itself might conceivably be aborted as being unnecessary.

Despite these developments, I would argue that development of the breeder continues to merit the strongest support. For each of the alternatives—lessened demand, additional uranium, "cows," solar, and pure fusion—is beset with inherent uncertainty. Demand for energy grows slowly now—do we *know* it will grow slowly after 2000 A.D.? Our estimates of uranium resources tend toward the high side now, but can we be confident of these estimates? Can we base energy policy on claims that uranium from the sea—or even from the Chattanooga shale—will be affordable? If uranium from the sea is not practical, are we prepared to face the environmental pollution caused by mining fuel for advanced converters, let alone the uncertain cost? Do we know that either the accelerator breeder or the fission–fusion breeder will work at a cost we can afford? And is the cost of solar electricity on a large scale likely ever to reach competitive levels? By contrast, the breeder is the only energy source that we know to be technically feasible that requires practically no mining and that can today produce unlimited amounts of electricity at a cost that is currently not more than twice the cost of electricity from coal or LWRs. In the bitter arguments that now rage over whether or not to go ahead with serious development of the breeder, this reality seems to be forgotten: the breeder eliminates *uncertainty* about the supply of uranium in relation to the demand for electricity in the future.

Thus, as many of us said many years ago, the breeder can confer a large measure of energy autarky on a country that can afford to use the breeder; and it is cheaper than solar electricity, and it avoids the profound technical uncertainties of fusion, the other two autarkic energy sources. For the United States, with its huge coal and uranium reserves, this may seem unimportant; but for France or Japan, both of which must import most of their energy, the prospect of energy autarky will always be attractive. That autarky extracts an economic penalty is perhaps not surprising. On the other hand, there is no law of nature that requires breeders always to be very expensive. Only by continued development—tedious and expensive though it may be—shall we be able to learn whether energy can be extracted economically from fully developed breeders.

There is another aspect of the breeder that has been insufficiently considered—namely, the possibility that a breeder, once built, will *never* be decommissioned. We have little understanding of what will limit the life of a LMFBR; indeed, there is probably little in such a plant that needs replacing and nothing that is not replaceable. Breeders (or, for that matter, even LWRs) may turn out to be much like dams: their lifetime is to be measured in centuries. Should this turn out to be the case, then one must recognize the possibility that the cost of energy in Harry Brown's reactor-dominated future could be far lower than it is today. Even to suggest such an eventuality must invite heavy criticism from skeptics; yet I cannot ignore that EBR-II has now been operating for 25 years, and there is *no* sign that it is wearing out.

VII

During my 42 years in nuclear energy, three events (other than the bomb) have stirred the strongest emotions in me: first, the achievement of the first chain reaction on December 2, 1942; second, my realization, in 1958, when I wrote "Energy as an Ultimate Raw Material; or Problems of Burning the Sea and Burning the Rocks,"[11] of how enormous was the store of energy in the low-grade ores of Phil Morrison and Harry Brown; and third, in 1982, as I stood in the control room of Phenix, the realization that a practical embodiment of inexhaustible nuclear energy was no longer merely a dream.

That the breeder is flawed because it is costly and may be misused to make plutonium for bombs must be conceded; but that these flaws cannot be overcome by human ingenuity I shall never concede. I remain confident that these deficiencies will yield to the imagination and courage of a new generation of nuclear technologists—a generation that surely will receive inspiration from Harry Brown's magnificent vision of a world of the future that draws much of its energy from common rock.

References

[1]H. G. Wells, *The World Set Free: A Story of Mankind* (Dutton, New York, 1914).

[2]H. Brown, J. Bonner, and J. Weir, *The Next Hundred Years* (Viking Press, New York, 1957).

[3]International Institute for Applied Systems Analysis, Wolf Haefele, program leader, *Energy in a Finite World: Paths to a Sustainable Future* (Ballinger, Cambridge, MA, 1981).

[4]H. Brown and L. T. Silver, "The Possibilities of Securing Long Range Supplies of Uranium, Thorium and Other Substances from Igneous Rocks," in *Proceedings of*

the International Conference on the Peaceful Uses of Atomic Energy, Vol. 8, Geneva, Switzerland, 1955.

[5]B. L. Cohen, "Breeder Reactors: A Renewable Energy Source," Am. J. Phys. **51**, 75–76 (1983).

[6]F. Daniels, *Direct Use of the Sun's Energy* (Ballantine Books, New York, 1964), p. 10. Daniels refers here to a symposium on solar energy held at the University of Wisconsin in 1953 and supported by the National Science Foundation.

[7]E. P. Wigner, "A Longer Range View of Nuclear Energy," in *Symmetries and Reflections* (Indiana University Press, Bloomington, 1967).

[8]W. B. Lewis, "Breeders are Not Necessary–A Competing Other Way for Nuclear Power," Atomic Energy of Canada, DM-69, AECL 1686, January 15, 1963.

[9]U.S. AEC, "Civilian Nuclear Power, A Report to the President," 1962.

[10]R. M. Rotty, A. M. Perry, and D. B. Reister, "Net Energy from Nuclear Power," Federal Energy Administration, FEA/B-76/702, 1976. Available from National Technical Information Service, Springfield, VA 22161.

[11]A. M. Weinberg, "Energy as an Ultimate Raw Material; or Problems of Burning the Sea and Burning the Rocks," Phys. Today **12** (11), 18–25 (1959).

Nuclear Power and Public Perception

Nuclear power is regarded by its proponents and by most nuclear experts as being necessary, acceptably safe, and desirable. It is regarded by skeptics and by a large fraction, if not a majority, of the American public as being unnecessary, unsafe, and undesirable. I shall refer to these two views as "the nuclear expert's view" and "the skeptical public's view," recognizing that not all who claim to be experts adhere to the expert's view and that a sizable fraction of the public do not accept the skeptical public's view. Because, in the democratic West, policy is ultimately based on public consent, unless the public accepts the expert's view of nuclear energy, the future of nuclear power is bleak.

The Nuclear Expert's View of Nuclear Power

The nuclear expert's view of nuclear energy, which is also the proponents' view, begins with the conviction that nuclear power is necessary. At the height of the oil crisis, few disputed this claim. Indeed, the 400-odd reactors in the world today (with a capacity of almost 275,000 megawatts), if replaced by oil-fired generators, would require some 6 million barrels of residual oil per day—some 10 percent of the world's total oil consumption. In the U.S. alone, were the approximately 105 reactors now on line to be replaced by oil-fired plants, they

---

Presented at the Yale School of Organization and Management, May 23, 1987, New Haven, CT.

would use about 2 million barrels of oil per day; by the year 2000, nuclear energy will have saved us some 7 to 12 billion barrels of oil! Of course, some of the plants actually replaced are coal-fired, not oil-fired, so these figures are rather exaggerated. Nevertheless, all would agree that the use of nuclear power has contributed significantly to the present oil glut and the consequent drop in the price of oil.

The need for nuclear power is reinforced by the extraordinary growth in demand for electricity. Although the United States used in total no more primary energy in 1986 than it did in 1976, the use of electricity in the United States increased by about 25 percent, during those ten years. This trend is discernible elsewhere also; the world is shifting to electricity even as it becomes, overall, more energy efficient. (I measure energy efficiency by the ratio of energy used to the gross national product in dollars. This ratio has fallen in the United States from 28×10^3 Btu per dollar of GNP in 1976 to 21×10^3 Btu per dollar of GNP in 1986.) As the demand for electricity grows, so will the pressure on all sources of electricity. Elimination of nuclear electricity means that more fossil fuel will be burned.

Nuclear energy is, on average, environmentally benign. This claim may sound absurd in the wake of Chernobyl; but even skeptics concede that *properly* operating nuclear power plants, together with *properly* operating subsystems, are environmentally benign. A nuclear power plant emits almost no radioactivity, although it adds more heat to the atmosphere than a coal-fired plant of equal capacity. By contrast, fossil-fuel plants emit carbon dioxide as well as acid-forming gases (sulfur dioxide and oxides of nitrogen) even when properly operating. Although, in principle, the acid emissions can be reduced, they can hardly be eliminated; and carbon dioxide cannot be eliminated even in principle, except by a vast reconstruction and reorganization of our fossil-fuel energy system.[1]

Accidents can happen. Nevertheless, at least for contained light water reactors, the probability of serious accident is small. The risk, defined as the product of probability and consequences,

$$\text{Risk} = \text{Probability} \times \text{Consequences},$$

is considered by nuclear experts to be acceptably small even though calamitous accidents cannot be totally ruled out.

The accident that shut down pressurized-water reactor number 2 at Three Mile Island (TMI-2) occurred after only 700 reactor-years of operation of light-water reactors (LWRs). This would suggest that the *a priori* probability of a serious core melt in such reactors is of the order of 1/700 per reactor-year. Since TMI-2, some 3500 LWR reactor-years have gone by without a core melt. A case can therefore be made for the core-melt probability being no larger than 1/3000 per reactor-year. The *a priori* core-melt probability, as computed by Rasmussen and by others,[2] is around 0.5×10^{-4} per reactor-year for the existing fleet of LWRs. With about 7,000 reactor-years of LWR operation worldwide expected since TMI-2 by the turn of the century, the *a priori* probability of avoiding a core melt by the year 2000 anywhere in the world is about 60 percent. Nuclear experts regard this as acceptable, since, as TMI-2 demonstrated, containment would probably prevent the spread of fission products. Moreover, for newer reactors, such as Sizewell B, calculated core-melt probabilities run in the range 10^{-5} to 10^{-6} per reactor-year.

At first sight, Chernobyl confounds such probabilistic considerations. The accident occurred after only 100 reactor-years of operation of Chernobyl-type reactors. Unlike TMI-2, in which less than one millionth of the inventory of dangerous fission products escaped containment, some 30 percent of the fission products escaped at Chernobyl. Thirty-one people died, some 200 suffered acute radiation sickness, and many millions received doses in excess of background. The Chernobyl reactor was moderated by graphite and cooled by boiling water. It was therefore very unlike a contained light-water reactor, such as TMI-2. Its positive void coefficient led to a violent nuclear excursion when the flow of cooling water was inadvertently interrupted. The disaster must therefore be attributed in good measure to Chernobyl's positive void coefficient. This design flaw is not shared by LWRs. Thus, the reactor at Chernobyl, being a faultily designed graphite reactor, has little bearing on the safety of contained, light-water reactors.

The disposal of high-level radioactive waste poses few technical problems. The high-level wastes are tied up in resistant matrices, either uranium dioxide, if fuel elements are not processed, or borosilicate glass, if the fuel is reprocessed. The ceramic or glass is encased in resistant cases and overpack. The entire package is sequestered in stable rocks or salt 1,000 or more feet below the Earth's surface. No one has imagined a sequence of events that could result in contamination of aquifers to a degree that could cause serious, large-scale damage to health. The radioactivity eventually decays; thus, radioactive wastes eventually are entirely benign, unlike industrial wastes, which often contain such stable poisonous elements as cadmium, arsenic, and lead. Radioactive wastes have already been sequestered, with considerable success, in the naturally occurring chain reactors in Gabon. Here, some 2 billion years ago, several tons of fission products and plutonium were generated in very rich bodies of uranium ore. French investigators have determined that the plutonium and most of the fission products remained at the site during the ensuing geologic eras, despite the long exposure to natural forces.

Proliferation of nuclear weapons is a hazard—but not one that can be attributed uniquely to nuclear power. All the overt nuclear weapons states manufactured their plutonium in simpler plutonium-producing reactors, not in reactors built originally for power. To be sure, plutonium can be manufactured in nuclear power plants, but this is a costly and somewhat inefficient way of making plutonium. In short, proliferation, being primarily a political problem, can only be solved by political means. Hanging nuclear power because it inevitably leads to proliferation of nuclear weapons is hanging nuclear power on a bum rap.

The Skeptic's View of Nuclear Power

Skeptics' objections to nuclear power often begin with a rejection of electricity as a desirable energy source. Electricity generation is inefficient, some two-thirds of the energy in the original fuel appearing as waste heat. Moreover, electricity generation tends to be centralized; for many skeptics, and for some of the public, centralization *per se* is regarded as socially undesirable. Only small is beautiful!

The growth in use of electricity is an aberration caused by the underpricing of electricity relative to other fuels. Were electricity priced at its marginal cost, the growth in its consumption would halt, and even be reversed. Beyond this, there are many new technologies for conserving electricity, which, as the price of electricity rises, will dampen its growth. Among these are the Philips high-efficiency electric bulb, variable-speed motors, and, eventually, high-temperature superconductors. Thus, few (if any) new central power plants of any kind will be needed. What additional electricity may be needed will be supplied by decentralized gas- or coal-fired cogenerators, or even windmills, or solar-powered photovoltaics.

Nuclear power is anything but environmentally benign: Chernobyl was the worst environmental disaster ever caused by a technological failure. Although Chernobyl was not a contained LWR, the violent explosion it underwent could, with some tiny probability, occur in a

LWR. For example, a catastrophic failure of the pressure vessel in a LWR cannot be assigned zero probability. Thus, the equation

$$\text{Risk} = \text{Probability} \times \text{Consequences}$$

does not adequately capture the impact of a calamitous happening. If the consequences are sufficiently dire, only *zero* probability is acceptable. On this ground alone, nuclear power must be judged unacceptable.

Although the wastes from nuclear power plants pose less chance of a calamitous accident, there is always some chance that water will penetrate the waste depository and that acquifers will be contaminated. To be sure, the dose to any individual as a result of such failure is small, but the number of individuals so exposed could be large. And if the health effects (predominantly cancer) can be expressed as

$$H = \alpha N \times D,$$

in which N is the number of individuals exposed to an average dose D in rems and α is the number of cancers per man-rem (generally taken to be 2×10^{-4} per man-rem), then the number of casualties induced even by doses of the order of background, 10^{-1} rem per year, could be large. For example, background imposes a dose of 25×10^6 man-rems on the U.S. population; at 1 cancer per 5,000 man-rems, this implies 5,000 cancers caused by natural background radiation each year. Of course, at this low dose, no one knows whether any harm is caused, but it is better to be safe than sorry.

Finally, although proliferation is primarily a political problem, why make proliferation any easier by spreading nuclear power plants to unreliable or belligerent states such as Iran, Iraq, and North Korea?

Much of the foregoing objection to nuclear power is beside the point; nuclear power in the United States has priced itself out of the market. No new nuclear power plant has been ordered in the United States for about 10 years, simply because nuclear power has turned out to be too expensive. With some power plants costing as much as $5,000 or more per kilowatt-hour of capacity, even solar energy begins to seem competitive. Thus, the market has put an end to the nuclear debate: nuclear electricity, being too expensive, at least in the United States, is dead and will probably not be resurrected.

Can the Experts' and the Skeptics' Views Be Reconciled?

The experts' and the skeptics' views of nuclear energy are starkly, even grotesquely, divergent. Though the general public does not accept all of the skeptics' arguments, especially with respect to the undesirability of electricity, most of the American public probably accepts most of the rest of the skeptics' view. Can we ever expect a sufficient consensus favoring nuclear power to develop?

In every country, public acceptance of the experts' view of nuclear energy seems to be well correlated with the underlying political structure.[3] In countries with an authoritarian tradition, the public's view and the expert's view diverge much less than in countries where the tradition is more democratic. Until the advent of *glasnost*, the Soviet Union developed nuclear energy aggressively, and the public's concerns were ignored. (The following anecdote is revealing. A few months after the Chernobyl accident, I spoke to a young engineering student in the Rossiya Hotel in Moscow. He quickly identified the radiation dosimeters in my breast pocket and explained that, in his high school course on civil defense, he had learned all about radiation meters. When I asked him whether he was worried about Chernobyl, he answered, "Not at all. The government and the Party take care of the people!") Of course, glasnost has changed all this: the Soviet public is now as anti-nuclear as the U.S. public.

France enjoys an atmosphere of acceptance of nuclear power. In part, this must be attributed to the extraordinary success of its nuclear enterprise: France has 49 operating reactors and 14 reactors under construction, and some 70 percent of its electricity was generated by nuclear reactors in 1986. This success I attribute, in turn, to the highly rational organization of the French nuclear program: a single electricity utility, EDF; a single supplier of equipment, Framatome, working in close concert with the Commissariat a l'Energie Atomique. This has resulted in fully standardized reactors. Because they are so similar, lessons learned at one reactor are quickly applied to other reactors. The French public seems to take great pride in these successes, especially since the view that nuclear energy is a French invention is widely held in France.

The French political tradition makes for more centralized, even authoritarian, government than the tradition in the U.S. or in other Western democracies. The Jacobins, after all, regarded themselves as an elite who interpreted the will of the people. Democracy in France has always retained this elitist, Jacobin tradition: French people seem remarkably content to delegate decisions to their elected leaders and to the governmental bureaucracy without perpetually second-guessing, even delegitimizing, their decisions. Thus, in a way, the French situation resembles that in the pre-glasnost Soviet Union: both are elitist, though of course to very different degrees. Nuclear energy seems to do best where the underlying political structure is elitist.

A third reason for the public's acceptance of the experts' view of nuclear energy in France is the lack of alternative sources of energy. Were it not for nuclear power, France would be depending very heavily on imported oil. Nuclear power has allowed France to reduce this dependence drastically, along with its consequent drain on foreign exchange.

Very different from France is the attitude toward nuclear energy in the United States. Here the immediate need for nuclear energy is harder to justify. The operating experience of American reactors has been less impressive than in France. Many partially built but now abandoned reactors burden both the landscape and utilities' balance sheets. A cumbersome licensing process adds to the cost and the time required to build reactors. Thus, the American public, which in 1975 answered *yes* by 3 to 1 to the question "In general, do you favor the building of more nuclear power plants in the United States?," answered *no* by 4 to 1 to the same question in May of 1986, shortly after the Chernobyl accident. Nevertheless, 75 percent of the public still believes that nuclear energy will be important in meeting future electricity needs.[5]

Public acceptance in other countries ranges widely—from Japan, South Korea, and Taiwan, where nuclear power is generally supported; to England and perhaps Belgium, where the public, especially after Chernobyl, is increasingly skeptical but the enterprise expands; to Germany and the Netherlands, where the future of nuclear energy hangs in the balance; to Sweden, Austria, Denmark, and Norway, where nuclear energy has been officially rejected.

One cannot discern a clear connection between the public acceptance and the state of the technology, nor even between public acceptance and the need for nuclear power. Sweden has no indigenous fossil fuels, and its 12 light-water reactors have been models of safe, efficient operation; yet, by referendum, Sweden is obliged to shut down all its reactors by 2010 A.D. (Since Chernobyl, a move has been under way to decommission the Barseback reactors, which are close to Denmark, sooner than 2010 A.D.) Japan, by contrast, has had an equally good operating experience, and it also has no fossil fuel, yet the public seems to accept nuclear energy, and Japan continues to expand its nuclear enterprise.

I have already suggested that the attitude of various publics toward nuclear power is rooted in the underlying social compact between the public and the ruling elite much more than is the objective performance of the nuclear enterprise. Where the tradition is participatory, open, and antiauthoritarian—as in the United States, the Scandinavian countries, Germany, and the Netherlands—the opposition to nuclear energy is powerful enough to all but stop the nuclear enterprise. Where the tradition is authoritarian, nonparticipatory, and closed—notably in the Soviet Union before glasnost, but also to some degree in France, South Korea, and Taiwan, and possibly even Japan—opposition to nuclear energy seems to be muted or even largely absent.

Approaches to Reconciliation

In a participatory democracy, such as the United States, public acceptance of a technology is necessary if the technology is to flourish. This means that licensing reform is no panacea for resolving the nuclear impasse. A technology cannot be regulated if it is regarded as an abomination by the majority, or even by a very vocal minority of the American people. Suppose commercial aviation were regarded by the public with the same suspicion with which it regards nuclear energy: Would not every judgment made, every license issued by the Federal Aviation Administration encounter the same intransigence that the actions of the Nuclear Regulatory Commission now encounter? And could air travel flourish under those circumstances? I believe not.

Similarly, the impasse in nuclear energy—indeed, the future of nuclear energy—rests upon the degree to which the public, particularly the articulate public, accepts the nuclear experts' view of nuclear energy as opposed to the skeptics' view. Should the public continue to regard nuclear energy with apprehension, distaste, and suspicion, nuclear energy will founder. What can be done, therefore, to bring the public's view more in line with the nuclear experts' view? Several possible approaches to such a reconciliation have been suggested; I review some of them in what follows.

Technical fixes

INHERENTLY SAFE REACTORS. David Lilienthal, the first chairman of the U.S. Atomic Energy Commission, argued in his book *Atomic Energy: A New Start* that nuclear energy could not survive unless it was based on reactors whose probability of catastrophic failure was zero.[6] At the time, Lilienthal was regarded as a nuclear skeptic; nevertheless, he expressed confidence that nuclear technologists could come up with reactors that were so obviously safe that even a skeptical public could be convinced of their safety. Lilienthal's challenge was at first greeted with skepticism by the technical community; but, some 8 years later, many ideas for inherently safe reactors (otherwise known as passively safe reactors) have been put forward.

Two major approaches to inherent safety have emerged: first, the modular high-temperature gas-cooled reactor, a graphite-moderated, gas-cooled reactor that is so small that, even in the event that it loses all of its cooling, the reactor does not melt; and, second, the Swedish PIUS reactor. This is a pressurized-water reactor at the bottom of a huge, water-filled, silo-like concrete pressure vessel. The concrete vessel is filled with water that contains boron (which quenches the chain reaction). During routine operation, two clever hydraulic locks separate the pure water that cools the reactor from the boronated water. Any upset in the flow of cooling water causes the locks to open and the boronated water to flood the reactor. The chain reaction stops and is cooled by natural convection for at least 10 days; during this time, one assumes a fire hose can be brought in if necessary to continue cooling after the original water in the concrete vessel has boiled off.

In a recent paper,[7] Lu Yingzhong, of the Institute of Nuclear Energy Technology at Tsinghua University in Beijing, has catalogued some dozen different proposals for inherently safe reactors. All of these proposed reactors depend for their safety not on complicated electrical, mechanical, and human interventions, but upon easily understood and passively embodied physical principles, such as gravity and thermodynamics. A 5-megawatt prototype of one such reactor is being built at Professor Lu's institute. China plans to deploy larger versions of such reactors to provide heat for Chinese cities. The reactor is an improved version of a system being built in the Soviet city of Gorki. (Incidentally, following Chernobyl, Andrei Sakharov announced his intention to devote himself to the design of inherently safe reactors.)

The reactors I have described differ radically from the light-water reactors that now dominate nuclear energy. Not surprisingly, most nuclear industry has rejected such drastic approaches; instead, several proposals for incrementally improved LWRs are being worked on in the United States and in Japan. These newer designs incorporate inherently safe features without being fully inherently safe. Safety still depends on timely mechanical interventions, which always have some small probability of failure. Probabilistic risk analysis suggests that the likelihood of a core melt in such reactors is perhaps 50 to 100 times smaller than in existing LWRs—say 10^{-6}/year instead of 0.5×10^{-4}/year.

WASTE DISPOSAL. Next to reactor safety, the problem of waste disposal evokes the shrillest objections to nuclear energy. This has always puzzled nuclear technologists since, technically, waste disposal seems to pose no insurmountable problems. Could it be that the public is simply suspicious of estimates based on analysis of the hydrologic and geologic conditions in a depository several thousand feet below the surface? After all, hydrology and geology are less exact sciences than physics and chemistry, and the public may sense this. This belief prompted Sweden, in response to its law requiring waste disposal to be "certainly" safe, to focus more heavily on developing a safe package and less on the geology of the disposal site. The Swedes have come up with a copper canister that all now agree can contain the wastes long enough for the dangerous elements to decay. The "perfect" container confers safety on the waste system by virtue of its chemical inertness, and finesses the need to know in detail the hydrology 2,000 feet below the surface. Thus, the "perfect" container is the analog for waste disposal of the inherently safe reactor. Apparently, the public understands and accepts this line of reasoning: waste disposal seems no longer to be a matter for major dispute in Sweden.

Should the United States devote more of its resources to developing an inherently safe container, rather than continuing to place primary reliance on geological security of the disposal site? This has been suggested by Luther Carter (1987) in his *Nuclear Imperatives & Public Trust: Dealing with Radioactive Waste.*[8] I believe it deserves serious consideration.

THE COST OF NUCLEAR POWER. New reactors are too expensive. Two approaches to reducing their cost can be identified. First, reactors can be simplified and perhaps reduced in size. Since no new reactor is being ordered in the United States, we cannot say how this will work here; on the other hand, current developments in Japanese–American advanced boiling-water and advanced pressurized-water reactors should lead to simpler and (one hopes) standardized designs.

Second, reactors probably will last longer than their design lifetimes; at any rate, refurbishing an old reactor is much cheaper than building a new reactor. Once a power plant has been paid for, the cost of electricity generated by the plant falls to just the operating and fuel costs. The cost of electricity then becomes quite small, since the fuel cost for nuclear plants is much lower than that for fossil-fuel plants. This trend toward lower costs as the plant ages is seen in many older nuclear plants. These plants are now gold mines for the utilities that own them, even when the expensive backfitting required by the Nuclear Regulatory Commission is taken into account.

Can technical fixes for perceived deficiencies in safety convince a skeptical public that nuclear energy is acceptable? We do not know. What is at issue is the public's acceptance of probabilistic arguments relating to safety. I sense that the public understands consequences, not probabilities. Even if probabilities are estimated to be very low, if consequences are large, the technology may be rejected. This is why I have been impressed by David Lilienthal's approach to inherent safety. Indeed, if the newer generation of reactors are sufficiently safe to be acceptable to skeptical experts—for example, the Union of Concerned Scientists or the Natural Resources Defense Council—then would not the skeptical public be convinced also? The public's skepticism has surely been catalyzed by these skeptical experts. Were they to say, "We now believe nuclear energy to be inherently safe," I should think the public would be strongly influenced by these views. Such an eventuality was hinted at by Jan Beyea, Chief Scientist of the Audubon Society. In testimony before Congress, he recommended that the United States launch a major development of both solar energy and inherently safe reactors. Should solar energy founder and inherently safe reactors prove practical, Dr. Beyea said that, in the twenty-first century, the environmental movement would then withdraw its objections to nuclear energy.

Social fixes

THE NUCLEAR REGULATORY COMMISSION. For many in the nuclear industry, the Nuclear Regulatory Commission has been largely responsible for the decline of nuclear energy in the United States. "Licensing reform," meaning a streamlining of the arduous process of approval now required by the NRC, has therefore become a catchword within the industry; and bills to achieve such reform have been before Congress for more than a decade.

Some interventions by the Nuclear Regulatory Commission have been ill considered. The most serious lapse was the NRC's closing of reactor number 1 at Three Mile Island (TMI-1) after the accident in reactor number 2 (TMI-2). TMI-1 was essentially intact: the decision to close it down was taken in the heat of concern and confusion at the time of the meltdown at TMI-2. Other NRC requirements—for example, the extraordinary measures now required to protect reactors against earthquakes—are probably counterproductive since the network of pipe supports often make routine maintenance difficult. In guarding against the small likelihood of an earthquake, NRC may have degraded the plant's resistance to more frequent challenges to its safety system.

But, overall, NRC has been indispensable. Just as a Federal Aviation Administration is necessary to assure the public that air travel is safe, so NRC is necessary to assure the public that nuclear energy is safe. Unfortunately, the public has greater confidence in the safety of air travel than it has in the safety of nuclear power. Under the circumstances, I would doubt that *any* changes in NRC will make enough difference to restore the public's confidence in nuclear energy in the short run.

What is needed most of all is accident-free operation of the existing nuclear system. NRC can contribute to achieving this goal, if by no other way than by heavily penalizing lax operators. With about 110 reactors operating in the United States, the *a priori* expected probability of a core melt in the United States between 1987 A.D. and 2000 A.D. is around 1 in 12. However, there are some 300 LWRs outside the United States; were their core-melt probability per reactor 0.5×10^{-4} per reactor-year, the probability of a core melt outside the United States might be around 1 in 4. A core melt anywhere is a core melt everywhere, as we witnessed in the aftermath of Chernobyl. Thus, even the best organized, most effective NRC can hardly affect the likelihood of accident outside the United States.

The incremental improvements in the technology of existing reactors, mandated by the NRC since TMI-2, may improve the outlook for accident-free operation in the United States. Each reactor in the United States now has been, or is being, subjected to a probabilistic risk analysis. In many cases, the resulting core-melt probabilities are lower than in the original Rasmussen analysis. Nuclear experts insist that, with each passing year, reactor operators learn more about possible weaknesses and how to correct them. Although it is fashionable to insist that the ways to catastrophe are innumerable, in fact they are, in any practical sense, finite. This is illustrated by the sequence at Crystal River, where a loss-of-cooling accident of the sort that devastated TMI-2 was aborted because the operators had learned how to handle the situation. Thus, one can make at least a plausible argument that we shall reach 2000 A.D.without another core melt.

THE STRUCTURE OF THE U.S. UTILITY INDUSTRY: NUCLEAR PARKS. In the United States, there are about 200 separate operating utilities. Of these, some 58 operate nuclear reactors. One utility, Commonwealth Edison, operates 12 reactors, while 25 utilities operate only one reactor apiece. Thus, the nuclear industry of the United States is fragmented and rather unstandardized. By contrast, there are only 11 utilities in Japan, and France has only one. We cannot prove that fragmentation necessarily leads to less safe operation, especially since the accident at TMI-2 occurred within a fragmented utility structure, and the accident at Chernobyl occurred in a centralized one. Nevertheless, strong theoretical arguments favor centralization as a means of avoiding accidents. Such centralization might take place in the United States if some of the hard-pressed smaller nuclear utilities are denied rate relief and go bankrupt. They presumably would be taken over by stronger neighboring utilities. This has not yet happened, but it remains a possibility.

The problem is, in good part, one of dissemination of information: ensuring that knowledge of near misses becomes common knowledge and that appropriate action is taken promptly. In typical American fashion, the utilities have not restructured into a few, more efficient operating entities; instead, they have set up an Institute for Nuclear Power Operation (INPO) as a clearing house, information exchange, and monitoring agency. INPO serves to improve communication among disparate operators of nuclear reactors, without requiring a complete restructuring of the industry into a few very large nuclear utilities.

Another structural approach that is not currently employed in the United States is the nuclear park. Half of the world's reactors are now sited in nuclear parks having four or more reactors. The largest such park, in Fukushima, Japan, has 10 reactors and generates almost 10,000 megawatts. Such parks, especially if the reactors in them are all the same, ought to possess much stronger cadres than are affordable at separated units. Learning ought to go faster there; evidence for this has been presented by Roberts and Burwell (1980).[9] They found that reactors (i.e., their operators) "learn" from their siblings on the same site faster than do similar isolated reactors. (The criterion of learning was the number of untoward events reported each year to NRC. In all cases, the number of such events diminished as the reactors aged, this being evidence of "learning." Reactors on multiple sites, however, "learned" faster than reactors on isolated sites.) Should utilities begin to build reactors again in the United States—say, after the turn of the century—it seems likely that the new reactors will simply be placed on existing sites. Of the 100-odd sites actually built or proposed in the United States, all but 20 are quite remote; an analysis by Burwell *et al.*[10] in 1979 suggested that the United States could accommodate a nuclear enterprise of 600 gigawatts without building any new sites.

COMPENSATION. Almost no one wants a nuclear reactor or a nuclear-waste facility in his vicinity (NIMBY—Not In My Back Yard!). As for reactors, the issue may be largely irrelevant, since most of the sites in the United States have already been chosen—they are the existing sites, most of which are remote and can be expanded. Not so with waste disposal. Despite the passage of the Nuclear Waste Policy Act of 1982, which calls for activation of a permanent geologic site by 1989, each state considered for a repository has raised serious objection to its being chosen. Since the act allows a state selected for a site to veto its selection, although the veto can be overridden by Congress, the waste issue can conceivably lead to a major constitutional impasse.

To forestall such confrontation between state and Federal government, massive compensation to the selected states is all but necessary. Senator Bennett Johnson, Chairman of the Senate Energy and Environment Committee, once proposed offering $100 million per year to the state selected for the repository and $50 million per year to the state selected for the Monitored Retrievable Storage Facility. This would add less than 0.2 mills per kilowatt-hour to the cost of nuclear power—less than 1 percent of its present cost.

Compensation may be necessary, but will it be sufficient to overcome the profound aversion much of the public now feels toward things nuclear? Compensation has been effective in Japan and in France; and even in the United States local host communities often benefit from the presence of large industrial enterprises. I have therefore been an enthusiastic advocate of compensation, particularly since the actual harm caused by a nuclear-waste repository is essentially zero, and that caused by a reactor, although probabilistic, is almost surely very small. Yet Governor Lamar Alexander of the state of

Tennessee opposed the Monitored Retrievable Storage Facility in Oak Ridge, even though he considered it safe, simply because of the negative perceptions its presence would evoke among prospective new industries in his state. To allay such perceptual anxieties, massive compensation, as suggested by Senator Johnson, ought to be effective.

Perceptual fixes: Low-level radiation

Of the two primary public worries about nuclear energy—fear of another Chernobyl and fear of radioactive wastes—I have argued that the first is a rational concern but the second is not. My recipe for dealing with future Chernobyls is to move to inherently safe reactors that, transparently and deterministically, once and for all, remove the specter of Chernobyl.

The concern about wastes, being less rational, is more difficult to deal with. Insofar as the worry is rational, it centers on the long-term effects of radiation at levels around background or less on large numbers of people. Actually, the biological effects of low levels of radiation are relevant even to the estimates of casualties in large reactor accidents. At Chernobyl, for example, the estimated population dose of perhaps 50,000,000 man-rems is delivered at an average dose about equal to background. If one assumes that 5,000 man-rems produces one cancer with no threshold and strict linearity between dose and response (the usual assumption), then one predicts 10,000 additional cancers.

But this is *not* a scientific estimate. As stated in the BEIR-III report on the biological effects of ionizing radiation,[11] science cannot say whether there are any deleterious effects of gamma radiation at levels around background. In any event, the effects, whether deleterious or beneficial, are surely very small.

Nuclear energy will always impose some small level of radiation on people. Unless and until the public at large accepts such very low levels of radiation insult as being insignificant, nuclear energy will always be faced with public skepticism. One can argue that such concerns about low levels of radiation are overblown, since the public accepts much larger radiation doses without any concern. Perhaps most notable is the radon in our homes. As Henry Hurwitz has often pointed out, the increased radon exposure caused by insulating all houses in the United States, imposes a cancer risk each year comparable to the cancer risk that would be caused were a major reactor accident to occur *every year*.

Will the public eventually get over its panic over low levels of radiation? Insofar as this panic is provoked by scientists who claim to know the etiology of cancer, I would say there is hope. As more is learned about cancer, it seems to me we are realizing that cancer, rather than being the penalty we pay for our polluting industrial civilization, is a part of the natural aging process: that whatever causes aging causes cancer. The most likely etiological agent is now believed to be the oxygen radicals produced as normal byproducts of metabolism. In short, we die eventually because we eat, not because pollution is doing us in. Perhaps as science strengthens the case for metabolism being the leading "cause" of cancer, and thereby weakens the case against low levels of radiation as a cause for cancer, the public's attitude toward low levels of radiation will ameliorate.

A Second Nuclear Era?

Will there be a second nuclear era? In the open societies, this will surely depend on whether the public can perceive nuclear energy as conferring benefits that outweigh the risks. This, of course, depends on how badly electricity from fission is needed, and this depends on what the alternatives are. Without belaboring the matter, I would argue that nuclear energy, via breeders, is the only energy source that is at once feasible and virtually infinite and that, in principle, could be made both inexpensive and extremely benign environmentally. In this, it is better than fusion (not feasible or, in any event, expensive), coal (carbon dioxide and acid rain), oil and gas (exhaustible), solar (too expensive and intermittent), and conservation (limited in impact).

In the short run, however, nuclear energy may founder in many countries—Sweden, Austria, and even the United States, blessed as we are with many other alternatives. But one must keep these matters in perspective. Eventually, unless solar energy or fusion can be made economical and practical, there will be no alternative except fission. Thus, in the very long run, I believe the answer must be yes, of course there will be a second nuclear era. The technology on which this second nuclear era will be based will almost surely be inherently safe—so inherently safe as to place relatively few demands on the operating skill and organizational integrity of the entities responsible for nuclear reactors and waste disposal facilities.

But I cannot say when the second nuclear era will begin. In France and Japan, the second nuclear era will probably merge smoothly with the current nuclear era. In the United States, it will come too, but only after the present generation of reactors have run successfully, the nuclear wastes have been sequestered, and newer reactors that the public perceives as being safe have been developed.

References

[1] *Energy in a Finite World, A Global Systems Analysis*, p. 804ff, W. Häfele, Program Leader, Ballinger, Cambridge, MA (1981).

[2] F. H. Fröhner, "Once Again, How Many Reactor Accidents," Nature **326**, 834 (1987).

[3] Michael T. Hatch, *Politics and Nuclear Energy: Energy Policy in Western Europe* (Univ. of Kentucky Press, 1986).

[4] David Marples, *Chernobyl & Nuclear Power in the U.S.S.R.* (MacMillan, London, 1987).

[5] U.S. Department of Energy, *Energy Security. A Report to the President of the United States* (U.S. GPO, 1987), p. 191.

[6] David Lilienthal, *Atomic Energy: A New Start* (Harper & Row, New York, 1980).

[7] Lu Yingzhong, "Ordeals of Chernobyl and the Re-Justification of Inherently Safe Reactors" (Working Paper, Institute for Energy Analysis, Oak Ridge Associated Universities, Oak Ridge, TN, 1987).

[8] Luther Carter, Issues in Science and Technology, Winter 1987, 46–61, 1987.

[9] P. C. Roberts and C. C. Burwell, *The Learning Function in Nuclear Reactor Operations and Its Implications for Siting Policy* [Institute for Energy Analysis, Research Memorandum #81-4(M), 1981].

[10] C. C. Burwell, M. I. Ohanian, and A. M. Weinberg, "A Siting Policy for an Acceptable Nuclear Future," Science **204**, 1043–1051, 1979.

[11] National Research Council, *The Effects on Populations of Exposure to Low Levels of Ionizing Radiation* (BEIR-III, 1980), especially page 5.

Engineering in an Age of Anxiety: The Search for Inherent Safety

Ours is an age of anxiety.* Although life expectancy (corrected for infant mortality) in the United States has increased by an astonishing 20 years since 1900, so has our fear of death. We no longer accept death, especially death from cancer, as "natural"—every death, even the inevitable deaths of old age, must be caused by some malign, and identifiable, outside influence. In this we seem to be reverting to a primitivism identified long ago by the French philosopher Lucien Lévy-Bruhl:[1] primitive societies do not accept natural death. Every death is attributed to a definite cause—if not to a hostile arrow, then to an enemy's malediction.

Modern technology is viewed, by and large, as a prime source of our anxiety. A succession of spectacular technological failures—Three Mile Island, Bhopal, Challenger, Chernobyl—has shaken the public's confidence in our engineering. And hardly a month goes by without one or another pesticide being proscribed and our fruits and vegetables being judged as tainted.

Some of our anxiety is justified. Bhopal and Chernobyl were disasters: no rationalization after the fact can undo the enormous loss in

---

Presented at Symposium on Engineering and Human Welfare, 25th Anniversary of the National Academy of Engineering, Washington, DC, 1989.

*Leonard Bernstein caught the essence of our times by entitling his Second Symphony "The Age of Anxiety."

public confidence that these incidents engendered. Moreover, our ever more interdependent technologies are more fragile because of their very complexity and reach: the Northeast power blackout of November 9, 1965, demonstrated to the public just how vulnerable a highly interconnected system is to inadvertent collapse. On the other hand, our preoccupation with manmade environmental insults at levels far below those imposed by nature's own pollutants is hardly rational.

All of this heightened concern, this dizzy oscillation from one fear to another, is amplified by our superefficient methods of mass communication. Had I the slightest idea of how to reduce the enormous amplification of public worry caused by television, had I a neat technological fix—an "inherently safe" television system that both guarantees our freedom yet avoids scaring us out of our wits—I would urge its development and adoption with the highest priority. Much of our efforts to develop inherently safe systems would then be unnecessary.

The Engineers' Response to Public Concern

These public concerns, both those that are justified and those that are not, place constraints on our technologies. Engineering always has required judgment concerning how safe is safe enough, and more safety can always be bought at a price. Where an engineer places his design point in a plot of "safety" versus "cost" is a judgment that is powerfully influenced by the public's concerns. This is illustrated by the difference between Western and pre-glasnost Soviet attitudes toward containment around pressurized-water reactors. Containment shells have been standard practice on Western commercial pressurized-water reactors since the first plant at Shippingport, Pennsylvania, went critical in 1957. Soviet PWRs were largely uncontained until Chernobyl. This difference was attributed to the West's being more sensitive to public opinion than the pre-glasnost Soviet Union. And of course the design point must always conform to regulatory standards, which themselves are sensitive to the public's concern. Thus, we read in the February 1987 report from EPA that "EPA's priorities appear more closely aligned with public opinion than with our estimated risks"[2]—an extraordinary admission that EPA, apparently at the behest of Congress, takes into consideration not only those public concerns justified by the facts but also those concerns that are not so justified.

I take as given, then, that the amount of safety a technology must provide is ultimately decided by the public; but the public is fickle and, at least as viewed by engineers, often ill-informed. Although public attitudes are distilled and codified through various regulatory bodies—EPA, NRC, FAA—conformance to standards promulgated by such bodies no longer guarantees acceptance by the public. Consider the Shoreham nuclear plant: the Nuclear Regulatory Commission has certified that Shoreham meets all regulatory standards—that is, "that there is reasonable assurance that it can be operated without undue risk to the health and safety of the public."[3] Yet the State of New York, perhaps more sensitive to the mood of the public than the NRC, has killed Shoreham.

Traditionally, engineers design devices to conform to safety standards established either by their own standards committees or by government regulators. Perhaps the first such standards were those imposed on boilers by the Engineering Insurance Companies of Manchester, England, following a series of boiler explosions about 100 years ago.[4] The worst such catastrophe was the explosion of the boiler on the Mississippi River boat *Sultana*, on 27 April 1865, with the loss of 1,500 lives. In the United States, the first American Society of Mechanical Engineers boiler code was adopted in 1914. Designers were required to incorporate safety factors—that is, if the bursting pressure of a vessel were P, the vessel was certified for operation at, say, $P/4$. Safety factors, as customarily formulated, were regarded as conferring "absolute" safety: a component designed and built in conformance with the appropriate code will not fail, provided approved operating conditions are not exceeded. Before the Age of Anxiety, the public trusted the engineer: devices designed according to code were "absolutely" safe—or at least they did not engender public apprehension.

Unfortunately, failures occur, even in components designed according to code—perhaps because operating conditions are exceeded through human error, or the component ages, or the component was defective in the first place—or even, perhaps, because the code itself may be defective. In instances in which failure can cause massive harm—as for example, in high-powered nuclear reactors—"probabilistic risk assessment" (PRA) has become an accepted tool for determining how safe a device actually is. In PRA, one tries to identify all the sequences of component failures that can lead to failure of the system as a whole. The components are generally designed to code; their failure rates, which theoretically are zero, must be estimated from historical data. For each sequence that leads to system failure, one estimates the "consequences" of that failure—the consequences may be loss of life, damage to equipment or to other property, ecological insult, and so forth. Thus, a probabilistic risk assessment results in a curve $P(C)$—that is, the probability of a failure that causes consequences of magnitude C as a function of C.

Probabilistic risk assessment became a recognized and accepted technique with Norman Rasmussen's study,[5] which was published in 1975. Rasmussen estimated that the median probability of a core melt for the Peach Bottom boiling water reactor was 3×10^{-5} per reactor-year, and for the Surry pressurized water reactor, 6×10^{-5} per reactor-year, the estimated probabilities being possibly too high or too low by a factor of five to ten. Most of these core melts would have few off-site consequences. The worst accidents, causing some 3,300 immediate radiation-induced deaths, 45,000 long-delayed extra cancers, and property damage of \$17 $\times 10^9$ was estimated to occur once every 10^9 reactor-years. (Chernobyl, an uncontained, very different reactor, caused only 31 immediate radiation casualties, but it did cause property damage of the order estimated by Rasmussen.)

Since the Three Mile Island accident, PRAs on 44 of the 111 power reactors in the United States have been completed, and an equal number are under way. Although the NRC does not quite use PRA as a regulatory standard, PRAs have been very useful in revealing vulnerable elements in a plant. For example, sequences involving small pipe ruptures generally contribute much more to the overall probability of core melt than do sequences involving larger failures. This has caused reactor engineers to focus much more on avoiding small "loss-of-cooling" accidents than had been the case before TMI-2. At that time—when, incidentally, PRA was not used widely—a popular view seemed to be that, if one takes care of the worst initiating event—say, a guillotine rupture of a large pipe—then the smaller initiating events would automatically be taken care of.

Curiously, even though PRA had its origins, even before Rasmussen, in the aerospace industry, it was not used by NASA. Indeed, ". . . early in the history of the Apollo program a decision was made not to use numerical probability analyses in NASA's decision-making process."[6] This rejection of PRA was justified on the grounds that PRA cannot substitute for human judgment. PRA simply offers an estimate of the probability of a failure, but it does not say how safe is safe enough. Were this estimate of failure very high—say, 10^{-1} per launch—a decision-maker could easily choose to abort. But ordinarily the probabilities of massive failure in a well-designed system are much lower—say, 10^{-3} or 10^{-4} per launch. Ultimately, therefore, the decision-maker must rely on his own judgment—that, say, 10^{-4} is acceptable, but 10^{-3} is unacceptable. In short, probabilistic risk assessment only provides options: for preassigned consequences, C, the probability of incurring C is $P(C)$. The engineer/regulator must choose whether or not the $P(C)$ curve characterizing a given system is acceptable.

Despite this shortcoming, PRA is now used rather widely in technologies that can pose serious hazard. NASA, at the behest of its post-*Challenger* hazard audit, has taken steps to incorporate PRA in its shuttle program. The Cabot Corporation, before embarking on a plan to bring liquid natural gas into Boston, conducted a PRA of the operation. AT&T has used PRA to analyze components of its communication system. After Bhopal, many chemical companies have used PRA to estimate the hazards of plants that handle toxic or explosive materials. And in nuclear industry, many countries that had not adopted PRA now make such analyses. It seems to have taken tragedies like Bhopal, *Challenger*, and Chernobyl to bring other technologies (or, in the case of nuclear power, other countries) to the same degree of sophisticated hazard analysis that has been employed by the nuclear industry in the United States for more than 15 years.

As I have already said, PRA does not relieve the engineer/regulator of deciding how safe is safe enough; it simply presents him with a quantitative estimate of the safety of a system. What can one do with such estimates, other than arbitrarily decide which system is safe enough and which is not? In short, how does one set safety standards? One possibility, and the one that has been adopted by the Nuclear Regulatory Commission, is to set the acceptable standards of safety at a level consistent with the standards established and accepted, *de jure* or *de facto*, for comparable technologies. Thus, the NRC, in its safety goals of 1986, accepted a societal risk from nuclear power no greater than the risk from other modes of generating electricity. This translates into a core-melt probability of 10^{-4} per reactor year; moreover, it accepts an additional risk of cancer from a nuclear plant no greater than one-thousandth of the natural risk. (Indeed, Howard Adler and I have argued for a general standard of radiation exposures to be the standard deviation of the natural background—that is, about 0.2 millisieverts per year.) Of course, any such judgment also has elements of arbitrariness: If the comparison is with a natural background of risks, what fraction of that background is acceptable? If the comparison is with other hazards incidental to a technology that achieves the same aim but that is already accepted, will the hazard of the new technology be accepted if it matches that of the accepted one? For example, dams fail at the rate of 10^{-4} to 10^{-5} per year. Does this mean that reactors should be accepted if their core-melt probability is also 10^{-4} to 10^{-5} per reactor-year?

The acceptable standard of risk depends on how widely the technology is deployed. As I mentioned in my essay "Salvaging the Atomic Age," this point was first stressed in 1963 by the Swedish aeronautical engineer Bo Lundberg.[7] He argued that, although the probability of a crash per passenger was independent of the total number of airplane flights, as the number of flights increased, so would the number of crashes in the entire system—*unless* the crash probability per flight were reduced as the system expanded. And if the number of crashes in the whole system became too great, the public would lose confidence in flying, even though the probability of any given passenger completing his journey without injury remained constant and at a level that was acceptable in 1963. As far as I know, this observation of Lundberg's was the first realization that acceptable standards of safety in our media-driven society are very sensitive to the public's perception.

The Search for Inherent Safety

Although PRA is now widely accepted by the engineering community as a means of assessing risk, I do not believe its use can allay the public's anxieties. The public understands consequences, not probabilities. To concede that a technology has the potential for causing a major disaster, even if the probability of such an occurrence is minute, is unacceptable in this age of anxiety. Faced with this stubborn reluctance by the public to accept probabilistic arguments, several technologies have sought to develop systems for which the risk, if not zero, is so low, and so transparently low, that the skeptical elites—nonestablishment experts who are opposed to nuclear energy—if not the public, will accept the technologies. This is the path urged by David Lilienthal, the first chairman of the Atomic Energy Commission. In his book, *Atomic Energy: A New Start*, which appeared in 1980, a year after TMI-2, Lilienthal urged nuclear technologists to redesign nuclear reactors so as to eliminate the chance of a repetition of TMI-2. In reviewing his book, I argued that a technical fix to achieve zero risk in a high-powered reactor was an oxymoron. Technologists could not eliminate the 15 billion curies in a 1,000-megawatt reactor nor the 200 megawatts of afterheat, which slowly decays, after the reactor has been shut off. The approach to zero risk had to be through better training and better institutions. Somewhat wistfully, I remarked at the seeming reversal of roles: the man of affairs, David Lilienthal, calling for a technical fix (inherent safety); the technologist, Alvin Weinberg, calling for a social fix (better training and better institutions).

This was not the first time the idea of an inherently safe reactor technology had been suggested. Peter Fortescue of General Atomics had been arguing in favor of "forgiving" nuclear reactors even before Lilienthal's book appeared. And five years before TMI-2, on June 1, 1974, a cyclohexane plant at Flixborough, England, blew up, killing 28 persons.[8] Had the explosion occurred during peak hours, many more would have perished. Soon afterward, Trevor Kletz, who was the safety adviser to ICI Petrochemical Division, coined the phrase "inherent safety" to connote chemical plants whose safety depended not on the active intervention of mechanical, electrical, or human agents but, rather, on the inherent characteristics of the process itself. Kletz became a proponent, if not a propagandist, for a new approach toward the design of chemical plants—an approach in which safety, rather than being achieved by clever devices and interventions *after* the basic process and plant layout has been chosen, becomes an integral part of the design from the very beginning. In supporting this new ethic of chemical-plant design, Kletz pointed out that changes in process had often been used in the past to bring the risk of at least certain accident sequences to zero. For example, thatch in new buildings in London was forbidden in 1212; airships now are filled with helium rather than hydrogen; and modern anesthetics and electric lighting eliminate the possibility of explosions from ether or poisoning from phosgene caused by the burning of chloroform in the gas lights used to illuminate operating rooms. Each of these process changes reduced to zero the risk of disaster from a particular accident sequence—but of course they did not eliminate all fires in houses, all airship disasters, or all deaths from overdoses of anesthetics.

Kletz used common sense in his approach to inherent safety in chemical plants. A primary principle was, *Never allow large accumulations of very toxic materials.* Had this principle been observed at Bhopal, the tragedy there would have been averted; the storage vessel that exploded contained around 15 tons of methyl isocyanate (MIC).

Modern plants that produce MIC no longer allow accumulations of more than a few kilograms.

Kletz has described several other common-sense steps that designers of hazardous chemical plants ought to observe—generally aiming toward replacing active systems (such as pumps) with passive systems (gravity) wherever possible; replacing highly exothermic reactions with less exothermic ones; but, above all, reducing inventory of hazardous materials.[9]

Although Kletz's views were quite well known among chemical engineers, I don't think inherent safety was taken entirely seriously in the chemical industry until Bhopal. A disaster of that magnitude is simply unacceptable. Will the mind-set of the chemical industry (which, until Bhopal, had not subjected its extremely hazardous operations, such as the production of MIC, to the same scrutiny the nuclear industry was obliged to accept) change enough to require inherent safety in its designs? As Robert Kennedy of Union Carbide said, "An aroused public can put us out of business just like it put the nuclear industry out of business."[10] Much activity within the chemical industry has now been directed toward designing and building safer plants. For example, the American Institute of Chemical Engineering, with broad support from the chemical industry, has established a Center for Chemical Process Safety. The Center, in some ways like the Nuclear Safety Analysis Center (which was established by the utility industry after the Three Mile Island accident), carries out studies and provides guidelines for improving the safety of chemical plants.

The strongest drive toward inherent safety has been in nuclear energy. From the earliest days, we realized that ours was a uniquely hazardous undertaking—the 15,000 million curies in a 1000-MWe reactor was an accumulation of radioactivity without precedent (and, incidentally, violated one of Kletz's dictums, keep inventories of hazardous materials low); moreover, the Hanford reactors, as built, had positive void coefficients* (though the Dupont Company eliminated the possibility of a catastrophic progressive flashing of steam and burnout of a single tube by taking most of the pressure drop across orifices placed at the inlet to each tube).[11] The choice of the Hanford site was itself an implicit recognition of a principle of inherent safety: place hazardous operations far from people. And SIR, the submarine intermediate reactor, near Schenectady, was housed in a containment

*A "positive void coefficient" means that, if the water used to cool the reactor flashes to steam (thus creating a "void"), the chain reaction intensifies.

sphere: an ultimate passive safety system that, it was believed, once and for all reduced the public risk to what the public would surely accept—*zero*. Thus the notion of inherent or, more accurately, passive safety in nuclear systems was to some degree implicit from the earliest days of nuclear energy.

But I do not think inherent safety was explicitly addressed in those days. After all, the PWR was an outgrowth of the nuclear submarine. A submarine reactor is designed primarily to be small and simple—this dictated the use of water both as coolant and as moderator. But compactness and, therefore, high power density, is incompatible with *inherent* safety. Thus, when Admiral Rickover's nuclear navy was moved to dry land with the building of a naval PWR at Shippingport, safety had to be provided by a complex array of active elements—emergency core-cooling systems, afterheat-removal systems, fast-acting shut-off systems, all of which require active mechanical and electrical interventions.

Three events brought inherent safety of nuclear plants to more prominence and explicitness: the rejection of nuclear energy in the Swedish referendum of 1980, Three Mile Island, and above all, Chernobyl. The two reactor accidents seem to have shown that some societies simply reject any probability, other than zero, of a reactor accident, especially one of the magnitude of Chernobyl. And the Swedish referendum, followed by *de jure* rejection of nuclear energy in Austria, Norway, Denmark, and, most recently, Italy, demonstrates how public concern is translated into public—that is, political—policy.

Kre Hannerz of the ASEA-Atom Company in Sweden was perhaps the first to recognize that a reactor for which the probability of a meltdown was essentially zero, if it could be devised, might restore the public's confidence in nuclear energy. He thereupon invented the Process Inherent Ultimately Safe (PIUS) Reactor. PIUS is a pressurized-water reactor in which the usual pressure vessel is replaced by a much larger prestressed concrete vessel. The reactor and its cooling system are immersed in a huge pool of water that contains boron. Should this borated water enter the core of the reactor, the chain reaction would stop since boron absorbs neutrons very strongly. The pure water that cools the reactor and the borated pool water are kept separate during normal operation by two open hydraulic density locks. As long as the system is operating normally, the boron remains outside; any upset (boiling, loss of coolant) automatically causes the borated water to enter the reactor without any mechanical device being actuated. The borated water bathes the reactor, shutting it down and providing automatic cooling for about a week with *no* external interventions whatsoever.

PIUS is, as far as I know, passively safe; that is, no credible sequence has been identified that could lead to meltdown, and no intervention—mechanical, electronic, or human—is needed to protect the system. Does this mean that PIUS is absolutely, or at least inherently, safe? If we equate inherently safe with *zero* probability of failure, I suppose the answer must be no. A small atomic bomb dropped on PIUS could cause a meltdown—as could, perhaps, an extremely powerful earthquake. But against lesser incursions—smaller earthquakes, conventional bombs planted by terrorists—PIUS appears to be secure. On the other hand, PIUS is surely *transparently* safe: the principle on which its safety depends is easily grasped.

The modular High Temperature Gas-Cooled Reactor (HTGR), first proposed by Reutler and Lohnert, represents another approach to passive, if not inherent, safety. This reactor is a more-or-less conventional helium-cooled, graphite-moderated reactor, in which the fuel elements are microspheres of uranium dioxide coated with essentially impervious layers of silicon carbide and pyrolitic graphite. The reactor is shaped like an elongated cylinder and produces only about 250 megawatts of heat. By virtue of its very high ratio of surface to volume and its low power density, the reactor can dissipate all afterheat passively to its surroundings without exceeding the temperature at which the microspheres release fission products. The modular HTGR (mod-HTGR) must therefore be regarded as inherently safe against such major challenges as those that caused the dramatic damage at TMI-2 and at Chernobyl. At the time of this writing, mod-HTGR is being considered by the Department of Energy as one of the reactors for producing tritium for thermonuclear weapons, and the Soviet Union and Germany are planning to build such a reactor at the Lenin Research Institute for Atomic Reactors in Dimitrovgrad.*

PIUS and mod-HTGR are the reactors closest to being inherently safe—that is, having no identifiable or credible failure modes. This statement is arguable: for example, is it impossible for air to enter HTGR and start a graphite fire? Or is it impossible to remove the boron from the large bath of borated water surrounding the PIUS reactor? Thus we see that inherent safety, in the sense of absolute

*Note added in 1991: Following Chernobyl, this project has been deferred.

safety, is itself hardly a credible concept when applied to large reactors. What is credible to a skeptic may be incredible to a designer. On the other hand, that both PIUS and mod-HTGR depend for their safety on passive, not active, systems is undeniable; and, perhaps even more important, their safety is relatively transparent and scrutable, in marked constrast with the safety of those reactors whose safety depends on quick intervention by many back-up auxiliary systems.

Between PIUS and mod-HTGR (whose safety depends almost entirely on passive or inherent elements) and existing reactors (whose safety depends mainly on active elements), there are now several reactors being proposed whose safety is improved by incorporation of several small changes on existing systems. I mention the metal-fueled sodium-cooled reactor, first proposed at Argonne National Laboratory. Because of the high heat conductivity of its fuel, this reactor is not subject to the loss of flow, the loss of external cooling, or transient overpower accidents of the standard LMFR. There are also the small advanced pressurized-water reactor of Westinghouse, and the BWR-600 of General Electric—essentially conventional light-water reactors in which the emergency cooling system is activated by gravity rather than by pumps. In the event of a loss-of-cooling accident, the emergency core-cooling system requires only the opening of a valve—an action that most would agree is certain.

The nuclear establishment on the whole dislikes the designation "inherently safe." The argument is severalfold. First, "inherently safe" cannot be equated with "absolutely safe"; no high powered reactor can under *every* circumstance, foreseen and unforeseen, avoid a meltdown. Second, if a reactor advertised as being inherently safe actually suffered a meltdown, the entire nuclear enterprise would suffer an intolerable loss of credibility. And third, can inherently safe reactors and actively safe reactors coexist? Would not the pressure to shut down existing, actively safe reactors become irresistible were newer, safer reactors to be built?

To these objections, I offer the following: I do not regard "inherently" safe and "absolutely" safe to be identical; inherent connotes to me only that the system's safety depends on inherent and passive characteristics. Under some all-but-incredible circumstances, an inherently safe system can cause harm—for example, a terrorist attack of sufficient strength. As for the failure of an inherently safe reactor compromising the future of nuclear energy, I think the failure of *any* reactor, inherently safe or otherwise, would be disastrous for the nuclear enterprise. And as for the coexistence of devices of different vintage and different degrees of passive safety, we already accept this in commercial air travel. Modern jet aircraft are more reliable than DC-3s, yet the public accepts both kinds of aircraft. Every technology improves, yet the introduction of a newer, safer technology generally does not cause the immediate shutdown of an older technology, provided the older technology has operated reliably and no better substitute for it is available.

How Safe Is Safe Enough?

I return to the central question: How safe is safe enough? From probabilistic risk analysis, we judge that the mean core-melt probability of the light-water reactors now deployed is between 10^{-4} and 10^{-5} per reactor-year. With about 400 reactors now deployed, the probability of a core melt between now and 2000 A.D. is between 4 percent and 40 percent. Many of us in the nuclear business are uncomfortable with this result, especially since a nuclear accident anywhere is a nuclear accident everywhere.

Incremental improvements and retrofits improve the odds. Thus, the core-melt probability for Sizewell-B, a large PWR under construction in England, and for the advanced PWR and BWR now being built in Japan, is between 1 and 5×10^{-6} reactor-years. In a fleet of 5,000 such reactors over their lifetime of 50 years, the probability of one such reactor suffering a core melt would be at least 25 percent. Is this good enough?

Having asked this question, I must confess that I cannot answer it—especially since the answer depends much more heavily on the public's perception of hazard than it does on the actual hazard. This is why I think inherent—or at least transparent—safety will remain a tantalizing goal for engineers in the coming century.

The public's views are, in this television age, strongly affected by skeptical and articulate elites. These are self-appointed spokesmen for the public interest as they conceive it. Though these elites are often antitechnological, many of them are sufficiently sophisticated to see trade-offs in any assessment of a technology's risk. Thus, Jan Beyea of the Audubon Society, shortly after Chernobyl, conceded before a Senate Committee that, in view of the warming produced by the greenhouse effect, nonfossil energy sources would be needed. Although solar energy was his favorite, he recognized that it was expensive. He therefore urged development of inherently safe nuclear energy. Beyea is not the only skeptic who has voiced this view: Paul Ehrlich, a strong opponent of nuclear energy, although espousing energy conservation as the inherently safe alternative to nuclear energy, has announced that, in view of the greenhouse effect, he would be prepared to accept nuclear energy, were inherently safe reactors to be developed.

So I think the nuclear community must address the skeptical elites. The engineering task is therefore to design reactors whose safety is so *transparent* that the skeptical elite is convinced and, through them, so is the public.

What the public requires by way of assurance will depend on the alternatives—and here the problem of the greenhouse effect may be decisive. I have estimated that, to defeat the greenhouse effect by the middle of the twenty-first century, we may have to deploy roughly 5,000 very large nuclear reactors. I assume here that total world energy use will increase to 500 quads per year from the present 300 quads per year; that the carbon dioxide produced by burning fossil fuels to the tune of 150 quads per year (75 million barrels of oil per day equivalent) will all be absorbed in the oceans; and that most of the remaining 350 quads will be supplied by many thousands of nuclear plants.

I have already pointed out that the probability of a single core melt in so large a fleet of reactors is only 0.005 per year, if the individual core-melt probability is 10^{-6} per reactor-year. This, however, leads to a probability of 25 percent over the 50-year life of the fleet. My own instinct, therefore, is to do better than this with transparently and passively safe reactors, rather than to depend entirely on incrementally improved existing reactors.

The notion of inherent safety seems to have political appeal. For example, the Italian Parliament, although putting an end to nuclear power in Italy after Chernobyl, has authorized a five-year study of inherently safe reactors as a possible way to bring nuclear back to life. Mikail Gorbachev has called for the development of inherently safe nuclear reactors, and the Soviet Union has offered to cooperate with other countries in their development.

Though the search for inherent—or at least passive—safety seems to be most evident in nuclear energy, the idea is catching on in the other technologies, and for pretty much the same reason. The state of New Jersey passed a Toxic Catastrophe Prevention Act in 1988 that places heavy restrictions on what kinds of chemical plants can be built there; and there is a chance that a federal toxic-catastrophe law may be enacted. Robert Kennedy's warning—that a skeptical public may strangle the chemical industry as it has the nuclear industry—must be taken seriously. Trevor Kletz's call for incorporating inherent safety wherever possible in all chemical plants makes good sense, whether or not the fate of the chemical industry is at stake.

Conclusion

William Clark reminded us, in *Witches, Floods & Wonder Drugs* (which I quoted earlier),[12] that much of today's anxiety over technology is as little justified as was the fear of witches that swept over Europe in the 15th, 16th, and 17th centuries. But people eventually overcame their fear of witches. According to Hugh Trevor-Roper,[13] the public relaxed after the intellectuals ("skeptical elites") of the time decided that witches were not as bad as their reputation, and then persuaded the public to regard witches as more or less benign. I do not say that today's technologies are always as benign as witches; the general anxiety, engendered in part by real catastrophes, in part by imagined catastrophes, has pushed the public's estimate of how safe is safe enough much further toward safety than most engineers would judge sufficient. But in this the public's voice is supreme; we as engineers have little choice but to respond to these concerns. Inherent safety, or at least transparent and passive safety, offers us a technical fix for the public's anxiety. I cannot guarantee that this technical fix will work, that the public will accept passively safe reactors, oil tankers, or chemical plants, when it rejects actively safe ones. That inherently safe (if not absolutely safe) technologies are feasible even in principle seems to me to be a remarkable existence theorem. I expect engineers of the coming generation will exercise their ingenuity, in the best engineering tradition, to reduce these exciting new ideas to commercial practice.

References

[1] L. Lévy-Bruhl, *La Mentalité Primitive* (F. Alcan, Paris, 1922). Also as described in M. Douglas and A. Wildavsky, *Risk & Culture* (University of California Press, Berkeley, 1982).

[2] Unfinished Business: A Comparative Assessment of Environmental Problems, Overview Report, Environmental Protection Agency Report PB 88-121048, p. xix.

[3] This is the statement that summarizes every approval of a reactor by the Advisory Committee on Reactor Safeguards.

[4] *Pressure Vessel Codes and Standards*, edited by R. W. Nichols (Elsevier, London, 1987).

[5] Reactor Safety Study: An Assessment of Risks in U.S. Commercial Nuclear Power Plants, NRC Report WASH-1400 (NUREG-75/014).

[6] *Post-Challenger Evaluation of Space Shuttle Risk Assessment and Management*, National Academy Press, January 1988, p. 55.

[7] Bo K. O. Lundberg, "Speed and Safety in Civil Aviation," The Aeronautical Research Institute of Sweden, Report 95, Stockholm, Sweden, 1963. Lundberg's estimates of the number of accidents per year in civil aviation and the goals he deemed necessary, follow:

Year	Estimated accidents per year	Goal
1980	120	~10
1990	250	
2000	440	
2010	710	~25

The actual number of accidents in commercial aviation in 1988 was 25.

[8] *The Flixborough Cyclohexane Disaster*, Report of the Court of Inquiry, R. J. Parker (chairman), Her Majesty's Stationery Office (1975).

[9] *Cheaper, Safer Plants or Wealth & Safety at Work*, Notes on Inherently Safer and Simpler Plants, the Institution of Chemical Engineers, 1985, p. 54.

[10] Wall Street Journal, September 20, 1988.

[11] From the very beginning, avoidance of criticality in handling of fissile materials was recognized as being essential. Codes governing the safe handling of such materials were developed largely through the efforts of D. Callihan and H. Paxton.

[12] W. C. Clark, *Witches, Floods & Wonder Drugs* (General Motors Symposium on Societal Risk Assessment, Warren, MI, 1979).

[13] H. R. Trevor-Roper, *The European Witch Craze* (Harper & Row, New York, 1969).

Nuclear Energy and the Greenhouse Effect

What is definitely known about the greenhouse effect, and what evokes argument among greenhouse experts?

What all agree upon is that the amount of carbon dioxide (CO_2) in the atmosphere has been increasing by more than one part per million (ppm) since 1958, when C. D. Keeling began monitoring CO_2 concentrations in the ambient air near Mauna Loa, Hawaii. At present, there are about 350 ppm of CO_2 in the atmosphere; this corresponds to about 750 gigatons of carbon (GtC) in the form of CO_2 in the atmosphere.

In the 1800s, before fossil fuels were used extensively, the concentration of CO_2 was only 280 ppm. For the past century, the amount of fossil fuel burned has increased each year, and it now stands at about 5.6 GtC/year.

During this century, many forests, including tropical forests, have been cleared. Although argument remains vigorous about how much CO_2 this process contributes, the number usually quoted is about 1–2 GtC/year; thus, between the burning of fossil fuels and the clearing of forests, a total of about 7 GtC/year, in the form of CO_2, is being injected into the atmosphere by human activities. The accumulation of carbon recorded each year in Keeling's measurements amounts to about 3 GtC/year. Thus, of the 7 GtC injected each year, only about

Presented at Midwest Energy Consortium Symposium, Chicago, IL, 1989; reprinted in Int. J. of Global Energy Issues **2** (3), 1990. Reprinted by permission.

half remains aloft (the so-called "airborne fraction," or ABF, is around 0.5). The other half goes, presumably, into the ocean and the biosphere.

In addition to CO_2, other greenhouse gases—methane, oxides of nitrogen, and chlorofluorocarbons—have been accumulating. Although the concentrations of these gases are several orders of magnitude lower than the concentration of CO_2, these gases are, per molecule, much more potent absorbers in the infrared than CO_2. This is partly because their absorption lines are stronger, but also, and more importantly, because the CO_2 spectrum is very heavily self-absorbed. As a result, the infrared absorption in CO_2 goes roughly as the logarithm of the concentration, whereas the infrared absorption in the trace gases is much more sensitive to their concentration.

Were there no greenhouse gases (of which water vapor is the most important), the average temperature of the Earth's surface would be 255 K. The greenhouse blanket deflects some of the outgoing radiation back to the Earth (roughly speaking), and this accounts for the observed temperature of 288 K. As evidence of the reality of greenhouse effects on planetary temperature, the extremely high surface temperature of Venus is generally attributed to the very high concentration of CO_2 in the Venusian atmosphere.

Many estimates of the magnitude of the greenhouse warming caused by a doubling of the CO_2 concentration—say, to $\sim$600 ppm—are now available. The usually quoted results lie in the range between $\Delta T = 2\,°C$ and $\Delta T = 5\,°C$ for a doubling of CO_2. On the whole, the various global circulation models predict different rainfall patterns.

The foregoing, I believe, pretty well summarizes what is known about the greenhouse effect. (I exaggerated; neither the contribution of forest clearing to the carbon budget nor the uncertainty in ΔT are known precisely. Indeed, there are some climatologists who insist that the "greenhouse effect" is a mirage.)

One of the controversial points is whether or not we are actually seeing a greenhouse effect in the very warm weather of the last few years. J. Hansen of the Goddard Space Institute has insisted that this is really the case.[1] On the other hand, Seitz, Jastrow, and Nierenberg[2] claim that at least some of the observed warming must be attributed to a fluctuation in the Sun's luminosity—a fluctuation not unlike the fluctuations observed in the luminosities of other G-type stars.

That the reality of a man-made greenhouse effect would be masked by natural climatic fluctuations was pointed out in 1975 by W. S. Broecker.[3] P. R. Bell has recently extended Broecker's analysis,[4] and I take the liberty, with Bell's permission, of presenting his result. Broecker and Bell begin with observations by Dansgaard *et al.*[5] on the Earth's temperature history as revealed by the oxygen-18/oxygen-16 ratios in Greenland ice cores. Broecker, using Dansgaard's superposition of two sinusoids of periods 179 and 82 years to reproduce the natural climate fluctuations, found evidence of a warming trend beginning around 1950.

Bell has now extended these results by identifying two additional periods, of 22 and 18.6 years. The 22- and 18.6-year periods are found in various other climatic records—for example, the Western High Plains drought record.[6] I find particularly convincing Bell's identification of these periods with known solar and lunar cycles; the 18.6-year period with the lunar nodal tidal cycle; the 22-year period with the Hale luminosity and sunspot cycle; the 179-year period with the beat of the lunar nodal precession and the lunar perigee period. Bell is less certain of the origin of the 82-year period.

Bell's analysis confirms as well as extends Broecker's findings—that, for whatever reason, the Earth's temperature has now diverged from the natural fluctuations (induced by solar and lunar variability) for the first time since 1851, a period of almost 140 years. I do not claim that this proves that greenhouse warming is here. I do assert that, since the fluctuations in solar luminosity are already accounted for in Bell's analysis, the warming seems to be *in addition* to whatever warming is induced by the solar fluctuations.

The Greenhouse Effect and Energy Policy

Let us assume, then, that the greenhouse threat is real and that the world ought to do what it can to reduce the accumulation of CO_2 and other trace gases in the atmosphere. I shall not speak further of the gases unrelated to energy production. Though these are important, their use (say, for refrigerants) is hardly essential, since substitutes are available, and their reduction hardly requires an excruciating choice between, say, living standard and global warming. Nor shall I speak much about reforestation, especially since there remains such controversy over the actual impact deforestation is making on atmospheric CO_2. Instead I shall confine myself to observations on the relation between energy policy and CO_2.

Three pieces of data are needed if we are to formulate a coherent policy on energy and CO_2 for the world:

TABLE 1. Carbon content of estimated recoverable reserve of fossil fuel.

	GtC	GtC in atmosphere, ABF = 0.5
Coal	3850	~1925
Oil (conventional)	130	~65
Gas (conventional)	120	~60

-Which fuels pose the greatest CO_2 threat?

-Which countries pose the greatest CO_2 threat?

-Is there an allowable greenhouse budget (AGB)—that is, an amount of CO_2 that, if injected, is completely absorbed in the ocean and the biosphere?

Let us consider each of these items in turn:

The relative importance of coal, oil, and gas

Per unit of energy, the amount of carbon released as CO_2 from coal is about 0.025 GtC per quad; from oil, 0.02 GtC per quad; from gas, 0.015 GtC per quad (probably an underestimate for gas, since use of gas entails the release of methane, which is, per atom, a much more potent infrared absorber than CO_2). Thus, a shift away from coal toward gas is CO_2 conserving.

The total recoverable resource of coal in the world is often estimated at 3850 GtC. Gas and oil recoverable from conventional resources are often put at one-tenth the recoverable coal reserve. Thus, we arrive at the following ultimate CO_2 burden resulting from the burning of *all* estimated (conventional) fossil fuel: note that I assume the airborne fraction remains at 0.5.

If these estimates of recoverable reserve can be trusted, we see that the greenhouse warming is primarily a coal problem: even if all of the conventional gas and oil were burned, the ambient CO_2 would not increase by more than about 20 percent. Of course, the nonconventional sources of hydrocarbons (oil shale, tar sands) redress the balance between coal and fluid hydrocarbons.

TABLE 2. Worldwide emissions of CO_2 carbon released in 1986.

Region	GtC	%	Cum. %
U.S.	1.36	24	24
U.S.S.R.	1.03	18	42
W. Europe	0.96	17	59
China	0.53	9	68
E. Europe	0.41	7	75
Japan	0.28	5	80
India	0.12	2	82
Canada	0.12	2	84
Others	0.94	16	100
World	5.7	100	100

Attributable to OECD = 50%.
Attributable to U.S.S.R. + E. Europe = 25%.

The offending countries

Table 2, taken from D. Bodansky,[7] lists the CO_2 emissions by country in 1986. Important to note is that fully 75 percent of the CO_2 comes from the industrialized world—indeed, from the part of the world that has the most experience with nuclear energy.

Is there an allowable greenhouse budget (AGB) other than zero?

Since only half of the emitted CO_2 appears in the atmosphere, the rest is being sequestered in the oceans and the biosphere. Firor, Harvey, and Maier-Reimer and Hasselmann[8] have independently suggested that, were emissions of manmade CO_2 to fall to about 3 GtC/year, all of the emitted CO_2 would be sequestered. Although this figure is disputed—for example, Emanuel at Oak Ridge National Laboratory claims (private communication) the AGB is only about 1 GtC/year—nevertheless, that there may be an AGB $\neq 0$ is extremely important, since this would set a limit on how much fossil fuel can be burned in a CO_2-conserving, or CO_2-static, world. I would suggest that determination of the AGB—through refined biogeochemical–oceanic modeling—ought to receive extremely high priority, comparable to that now given to general circulation models of the climate.

What's to Be Done?

There are many, notably the Harvard economist Tom Schelling, who insist that nothing really can be done to stave off greenhouse warming and that the only realistic and prudent policy is mitigation and adaptation. Schelling may be right, yet such counsel of despair seems to many of us to be inconsistent with our confidence in our ability to control our destiny.

What, then, is to be done? Or, more accurately, are there energy policies that would mitigate greenhouse warming but that make good sense whether or not the greenhouse effect proves to be real?

On one element of policy, all will agree: *Conservation*. That energy can be decoupled from the GNP has been demonstrated by the U.S. experience. We used no more energy in 1986 than we did in 1973, even though real GNP increased by 40 percent. On the other hand, electricity increased at almost the same rate as the GNP in that period. The constancy of energy consumption represents a remarkable departure from the trends of the previous decades, when energy was increasing at 4 percent per year. Almost every energy prognosticator was proven to be a fool: the last decade has demonstrated that conservation—meaning a continued reduction in the ratio of energy to GNP (E/GNP)—has been an extraordinary success.

Most explanations of the success of conservation rest on the working of the market: as energy becomes expensive, individual users conserve. Other factors included a shift to service and away from manufacturing, and the influence of a widespread conservation ethic.

A factor that several analysts stress (C. Burwell, W. Devine, S. Schurr[9]) is the growth in the use of electricity. Electricity, though inefficient at the point of generation, is exquisitely efficient at the point of use: the overall effect is that substitution of electricity in many processes saves *primary* energy.

I mention these alternative explanations of the improvement in the E/GNP ratio to suggest that conservation is a complicated, and not completely understood strategy. In particular, I am unpersuaded that, in every instance, conservation is a better bet than an increase in supply. To be sure, in the United States we have used 26 quads less energy in 1986 than we would have had the E/GNP ratio remained at the 27×10^3 Btu per 1982 dollar of GNP instead of the 21×10^3 Btu per 1982 dollar of GNP realized in 1986. Assuming the savings came 50 percent from oil, 25 percent from coal, and 25 percent from gas, I conclude that conservation, whatever its origin, has allowed the

TABLE 3. Energy and carbon-emission profile for France.

	1978		1987	
Source	Quads	GtC	Quads	GtC
Coal	1.21	0.030	0.69	0.018
Oil	4.72	0.094	3.42	0.068
Gas	0.83	0.012	0.99	0.015
Hydro	0.60	--	0.58	--
Nuclear	0.25	--	2.12	--
Total	7.61	0.136	7.80	0.101

United States to emit 0.5 GtC less in 1986 than it would have had our economy not become more efficient. This is to be contrasted with the 0.1 GtC saved in 1986 by the operation of the 100 nuclear reactors in the United States. Such comparisons have sometimes been used to deny nuclear energy *any* role in reducing CO_2 emissions: a dollar spent on conservation is claimed always to save much more CO_2 than a dollar spent on nuclear energy. Although this may be true at present, where we are squeezing fat out of our energy economy, it is by no means clearly the case after the "easy" conservation has been achieved, or after nuclear energy has been rationalized, as it has in France. In France, for example, nuclear energy has played a central role in reducing CO_2, as seen by comparing the CO_2 emissions in France in 1978 and 1987 (Table 3). Although France's total energy increased by 2.5 percent in this 10-year period, the CO_2 emissions fell by 26 percent!

A Nuclear Path to Salvation from CO_2

Can nuclear energy play a role in reducing CO_2 emissions on a worldwide scale by, say, the year 2040—50 years from now? I think the answer is yes, but the path will be strewn with difficulties. First, we must decide how big an energy system the world might require in 2040 A.D. The current world energy output is about 300 quads, corresponding to a carbon emission of 6 GtC/year. For the year 2040, I choose, rather arbitrarily, an energy budget of 500 quads. Such a budget already implies very stringent conservation; other estimates (see "Energy in Retrospect") have ranged from a high of 1,000 quads in 2030 A.D. to a low of around 320 quads in 2020 A.D.

I shall now assume the AGB is 150 quads, corresponding to a yearly release of 3 GtC. This leaves an energy budget of about 350 quads to be supplied by nonfossil sources: solar (including wind, hydroelectric, biomass conversion, and direct solar), geothermal, fusion, and fission. Let me assume that all nonfission, nonfossil sources supply 50 quads per year. This leaves 300 quads of primary energy to be supplied by nuclear plants. Is this nonsense?

A 1,000-GWe nuclear electric power plant utilizes about 0.06 quads per year of primary energy. To supply our 300 quads of primary energy would require the equivalent of 5,000 1-GWe power plants—some 10 times as many as now are operating or are under construction. In such a world, electricity, which now accounts for about 30 percent of the world's primary energy, would increase to 60 percent. This in itself is not so daunting, inasmuch as the world has been electrifying for many decades. In the United States, electricity in 1968 used 18 percent of our primary energy. Today, it accounts for 36 percent, and there is little sign that demand for electricity is abating.

If reactors are to be built on a large scale, their price will have to come down—to the cost in France, about \$1,500/kWe rather than the cost in the United States, which is almost twice that value. At \$1,500/kWe, or \$1.5 $\times$ 10^9/GWe, the entire fleet would cost about \$7.5 $\times$ 10^{12}. Spread uniformly over 50 years, this amounts to a yearly expenditure of \$150 $\times$ 10^9—large, but not intolerable, considering that the world's military budget is six times as much.

I would visualize this huge deployment of nuclear energy to be largely confined to the OECD and Eastern bloc, both of which possess ample nuclear know-how, are relatively rich, and are currently the largest producers of CO_2. The AGB of 150 quads would be largely allocated to the developing, nonnuclear world.

A world of 5,000 large reactors would use about 0.75 $\times$ 10^6 tons of uranium each year if the reactors were all no-recycle LWRs, 0.4 $\times$ 10^6 tons per year if they were all recycling LWRs, and only 0.015 $\times$ 10^6 tons per year if they were all breeders. I compare these figures with the world's uranium reserve at a price below \$130/kg estimated in 1983 by the Nuclear Energy Agency of OECD (Table 4). If we assume that the speculative resources will materialize, for a total of 31 $\times$ 10^6 tons world-wide at less than \$130/kg, a 5,000-reactor fleet could be sustained for about 40 years without recycling; for about 80

TABLE 4. Estimated uranium resources ($\times 10^6$ metric tons).

	Non-Communist world	Communist world	Total
Reasonably assured resources (RAR), < \$130/kg U	2.0	1.0(?)	3.0
Estimated additional resources (EAR), < \$130/kg U	2.3	1.2(?)	3.4
Speculative resources (SR), < \$130/kg U	6.3–16.2	3.2–8.1	9.5–24.3

Source: OECD/NEA, Uranium: Resources, Production and Demand, December 1983.

years with recycling; and for about 2,000 years with breeders. (These estimates may grossly underestimate the time at which breeders might be needed.) Let us assume that breeders might cost \$300/kWe more than LWRs. To compensate for a \$300/kW higher cost, I have estimated[10] that the price of a kilogram of uranium would have to rise to around \$500. The amount of uranium available at this price is enormous; and the added cost of electricity, if such expensive uranium were used, would be only about \$0.01/kWh. Thus, I would conclude that a 5,000-reactor world, even based on nonrecycling LWRs, could probably be fueled for several centuries and possibly much longer.

Recently, N. Mortimer of the Friends of the Earth[11] has stressed that nuclear plants use fossil fuel during construction and for preparation of fuel and therefore contribute to the accumulation of CO_2. This matter was examined by Rotty, Perry, and Reister[12] at the Institute for Energy Analysis. The least favorable case was a 1,000-MW(e) PWR with no recycling, using uranium from Chattanooga shale enriched from its natural content of 0.7 percent ^{235}U to 1.0 percent ^{235}U, producing, over its 30 years of operation, 197,100,000 MWh of electricity. The results are given in Table 5. The electricity required, 13,750,000 MWh, requires, at 10,000 Btu (of coal) per kilowatt-hour, 1.38×10^{14} Btu, which releases 3.5×10^{-3} GtC; the 1.51×10^{14} Btu of fuel (oil) releases 3×10^{-3} GtC, for a total carbon release of around 6.5×10^{-3} GtC over the 30-year life of the plant. Had the plant been coal-fired, it would have released 0.05 GtC; thus the nuclear plant, over its lifetime, releases only 13 percent as much carbon as would a coal-fired plant of the same capacity. For the breeder, the

TABLE 5. Lifetime energy requirement for a 1,000-MW(e) pressurized-water reactor with no recycling, 0.30% enrichment of Chattanooga shale, producing an output of 197,100,000 MWh.

Process	Electricity required (MWh)	Fuels required[a] ($\times 10^6$ Btu)
Mining 5,682 metric tons natural uranium	1,667,000	20,010,000
Milling 5,682 metric tons natural uranium	2,736,000	99,800,000
Conversion of 5,682 metric tons natural uranium	82,960	7,676,000
Enrichment of 3,022 $\times$ 10^3 separative work units (kg)	8,533,000	2,412,000
Fuel fabrication of 822 metric tons enriched uranium	247,400	2,109,000
Power plant: construction plus 30 years of operation	461,500	18,140,000
Reprocessing[b] of 822 metric tons of fuel	16,360	292,600
Waste storage: 30 years of operation	5,010	183,200
Transportation:		
5,682 metric tons of natural uranium	597	81,930
822 metric tons of spent fuel	1,861	255,900
Total required energy	13,750,000	151,000,000

Source: Perry, Rotty, and Reister (Ref. 12).
[a]Not including any fuels used to generate electricity.
[b]Without recycling of either uranium or plutonium, there probably would be no reprocessing. On the other hand, we have not evaluated the energy required for storage of spent fuel. Since this is a very small item, we allow the reprocessing and waste storage to stand as an estimate for the unevaluated storage of spent fuel.

release is much less. Thus, a 5,000-reactor world based on no-recycle PWRs and fueled by uranium from Chattanooga shale would release 33 GtC into the atmosphere over 30 years, or about 1.1 GtC/year. This is about 40 percent of the 3 GtC/year thrown into the atmosphere by the 150 quads of fossil fuel assumed in our scenario, which is certainly not negligible. However, these estimates are probably too high, since newer methods of isotope enrichment are much more energy efficient than is gaseous diffusion. One conclusion I draw is that, even if economics does not favor breeders, reduction of CO_2 emission does favor their introduction.

Finally, I must speak of proliferation, which, in the view of many skeptics, is the ultimate insoluble problem of nuclear energy. Given that any competent state intent on developing a bomb can eventually do so, with or without nuclear power, is the danger of proliferation so overwhelming as to force the world to reject nuclear power, even in the face of CO_2-induced global warming? Those who are unwilling to reject nuclear energy stress, first, that the connection between nuclear power and proliferation is not nearly as close as the skeptics believe[13] and, second, that certain modifications in siting and reprocessing policy could further break the connection between nuclear power and nuclear proliferation. A proliferation-resistant siting policy would concentrate the world's 5,000 reactors on no more than 500 sites. (In the United States, there are already about 75 sites; in the entire world, around 235.) Each site would house about 10 reactors. Reprocessing would be conducted only on these sites and under close international supervision. Confined siting and internationally supervised reprocessing tend to make proliferation more difficult—although, as I have already concluded, proliferation is in some sense an insoluble problem. Technical methods can complicate unauthorized diversions of plutonium, but, ultimately, the world must depend upon political resolution of this issue. At the moment, we seem to be in a great political transition whose ultimate outcome—ending the cold war and laying the ground work for Immanuel Kant's comity of liberal democracies—may make proliferation less troublesome than it now seems to be.

I cannot say what conditions are sufficient to allow such revolutionary nuclearization. It is much easier to say what conditions are necessary. These include (1) accident-free operation of existing reactors; (2) reduction in cost of nuclear reactors—say, to the \$1,500/kWe achieved by French reactor builders; (3) acceptable sequestering of wastes; (4) reduction in core-melt probabilities and/or development of passively safe (inherently safe) reactors.

Of these items, I regard item 1 as being most important in the short run and item 4 as being most important in the long run. At present, the core-melt probability for a modern PWR, as estimated by probabilistic risk analysis, is around 10^{-4} per reactor-year. Were 5,000 such reactors deployed worldwide, the estimated core-melt frequency for the entire fleet would be one every two years. This I regard as being unacceptable. On the other hand, I cannot, nor can anyone, specify exactly what core-melt frequency would be acceptable. It is for this reason that I favor so strongly the growing efforts to develop passively safe or inherently safe reactors—such as Mod-HTGR, PIUS, and metallic fueled LMRs. Core-melt probability for reactors of this type is estimated to be extremely small. My own hope is that the safety of such systems will be so transparent that the skeptical elite and, through them, the public can be convinced that whatever risks these devices pose are acceptable. For unless the public regains confidence in nuclear power, this option will remain foreclosed.

Skeptics may argue that all this is unnecessary—that solar energy will be a better buy within, say, the next 15 years; or that fusion will work; or that even my estimate of 500 quads of energy per year by 2040 A.D. is too high. All these things may be true, but we cannot depend upon their happening. It seems to me, therefore, that simple prudence suggests fixing nuclear energy along the aforementioned lines. If greenhouse warming turns out to be real, nuclear energy can then be one of the supply options that we will turn to. If greenhouse warming turns out to be fictitious, an acceptable, economic nuclear option is surely better than one that does nothing but create strife and dissension. If the other nonfossil supply options—solar, fusion, geothermal—prove to be cheaper or more acceptable than nuclear, they will relegate fission to an insignificant position, one with which even nuclear enthusiasts like me would not quarrel. But let us remember that nuclear energy is the only nonfossil source, other than hydroelectric, that has been demonstrated to be practical and economical, where it has been done right (in France and in Japan). To deny it a role in the formidable task of controlling CO_2 is surely shortsighted.

References

[1] J. Hansen, I. Fung, A. Lacis, D. Rind, S. Lebedeff, R. Ruedy, and G. Russell, "Global Climate Changes as Forecast by Goddard Institute for Space Studies Three-Dimensional Model," Journal of Geophysical Research **93**, 9341–9364 (1988).

[2] F. Seitz, R. Jastrow, and W. Nierenberg, *Scientific Perspectives on the Greenhouse Problem* (George C. Marshall Institute, Washington, DC, 1989).

[3] W. S. Broecker, "Climate Change: Are We on the Brink of a Pronounced Global Warming?," *Science* **189**, 460–463 (1975).

[4] P. R. Bell, "Reprise and Extension of Broecker's Prediction: Factors in Hemispheric Temperature Records" (Institute for Energy Analysis, Oak Ridge, TN, 1989).

[5] W. S. Dansgard, S. J. Johnson, H. B. Clausen, and C. C. Longway, Jr., "Climatic Record Revealed by the Camp Century Ice Core," in *Late Cenozoic Glacial Ages,* edited by K. K. Turekian (Yale University, New Haven, 1971).

[6] P. R. Bell, "The Combined Solar and Tidal Influence in Climate; Predominant Periods in the Time Series of Drought Area Index for Western High Plains, 1700–1962," Proceedings of Workshop at Goddard Space Flight Center, Greenbelt, MD, 1980, pp. 241–264.

[7] D. Bodansky, "Nuclear Power in the Context of Global Warming," presented at Conference on Cooperative Measures for Reducing Global Tensions (Los Alamos National Laboratory, Sante Fe, NM, 1989).

[8] J. Firor, "Public Policy and the Airborne Fraction," Climate Change **12**, 103–106 (1988); L. D. Harvey, "Transient climatic response to an increase of greenhouse gases," Climate Changes **15** (1,2), pp. 15–30 (1989); E. Maier-Reimer and K. Hasselmann, "Transport and Storage of CO_2 in the Ocean: An Inorganic Ocean-Circulation Carbon Cycle Model," *ibid.* **2**, 63–90 (1987).

[9] The work of these authors is summarized in EPRI Journal, Dec. 1989: "Electricity and Productivity"; also, S. H. Schurr, C. C. Burwell, W. D. Devine, Jr., and S. Sonemblum, *Electricity in the American Economy* (Greenwood Press, New York, 1990).

[10] A. M. Weinberg, "Nuclear Energy and Proliferation: A Longer Perspective," in *The Nuclear Connection,* edited by A. M. Weinberg, M. Alonso, and J. Barkenbus (Paragon House, New York, 1985).

[11] N. Mortimer (1989), "Proof of Evidence, Aspects of Greenhouse Effect," Friends of the Earth Ltd.

[12] A. M. Perry, R. M. Rotty, and D. B. Reister, "Net energy from Nuclear Power," in *Nuclear Power and Its Fuel Cycle* (International Atomic Energy Agency, Vienna, 1977), Vol. 1; D. Spreng, *Net-Energy Analysis and the Energy Requirements of Energy Systems* (Praeger, New York, 1988).

[13] *The Nuclear Connection, op. cit.*

The First and Second Fifty Years of Nuclear Fission

By 1944, the Chicago Metallurgical Laboratory had pretty well completed its design of the Hanford graphite-moderated reactors, and so much of our attention was turned to future applications of nuclear fission. We organized a New Piles Committee, which met weekly for three months during the spring of 1944. Here the senior luminaries, such as Fermi, Wigner, Szilard, and Franck, together with me and a few other younger assistants, discussed various ideas for reactors: for power, for submarines, for production of plutonium, even for inducing endothermic chemical reactions. Our imaginations ranged widely, as we considered various moderators, coolants, and configurations. Inventing a new reactor was an everyday occurrence, simply because, as far as we knew, no one else had thought about these matters.*

At that time, we were under the impression that uranium would always be scarce. Our thinking was, therefore, directed mostly to breeders: unless breeders were successful, nuclear energy would not survive very long. Of course, this all happened almost twenty years before Bennett Lewis thundered in his famous paper, "Breeders Are

Based on a paper presented to Conference on Nuclear Fission: Halfway to the First Century, Ottawa, Canada, 1989. This version presented at Conference on Technologies for a Greenhouse-Constrained Society, Oak Ridge, TN, 1991.

*Bertrand Goldshmidt in his recent book, *Atomic Rivals*, mentions that Hans Von Halban, in 1940, had patented a reactor moderated and cooled with light water.

TABLE 1. World nuclear power plants by type, December 31, 1990.*

Reactor type	Number of units	Total output (GWe)
PWR	306	268
BWR	100	83
GCR	39	13
HWR	50	26
LGR	20	16
LMFBR	10	5
Operable	413	318
Under construction	112	93
Total	525	412

*Source: *Nuclear News*, February 1991.

Not Necessary," that there are ways of skinning the cat of uranium scarcity short of full-fledged breeders. This question, in 1991, still remains moot!

The Main Lines of Reactor Development

Although at least 20 reactor concepts have received serious consideration since 1944, only five types have become commercial power plants: light water (in both pressurized and boiling-water versions); heavy water (CANDU); graphite-moderated, steam-cooled (RBMK), like the reactor at Chernobyl; gas-cooled graphite; and liquid-metal cooled fast breeders. The world's total of 525 nuclear reactors in operation and under construction in 1990 have an electrical capacity of 412 GWe, or about 10 percent of the world's total installed electrical capacity. The electrical energy produced in 1990 by these reactors was more than 17 percent of the world's electricity and about 6 percent of the world's primary energy. This widespread use of fission as a source for 17 percent of the world's electrical energy must be regarded as an extraordinary achievement.

Of the world's 500-odd commercial reactors, about 85 percent are moderated and cooled by light water. As one who was involved in the original decision to power *Nautilus* with a light-water reactor, I have never outgrown my astonishment that LWR became the dominant reactor type. After all, light water was chosen originally for the submarine because such reactors are compact and, at least in principle, relatively simple, not because light water recommended itself as the best choice for generating electricity cheaply. To achieve compactness, we had to use highly enriched uranium, and, at the time, enriched uranium was very rare and expensive. Moreover, to retain the simplicity of its core design, the original *Nautilus* prototype simply burned ^{235}U, without generating any new fissile material. Given our impression at the time that enriched uranium was very scarce, we could not visualize a light-water reactor ever producing electricity at a competitive cost.

Several developments changed our perception. The first was the investigation of natural uranium lattices moderated by ordinary water. These exponential experiments were carried out on top of the X-10 reactor in Oak Ridge during 1944. The experiments were motivated by calculations that I had performed, at Eugene Wigner's suggestion, on natural uranium lattices moderated with H_2O. Bob Christy collaborated on the earliest calculations, and he and I patented a "Light Water Moderated Neutronic Reactor," U.S. Patent 2,806,819, filed January 9, 1947 and issued September 17, 1957. In this patent, we estimated that uranium enriched in ^{235}U to only 0.75 percent (compared to the natural 0.71 percent) would chain react.

At Oak Ridge, some 18 separate exponential experiments were performed. The best value of k was tantalizingly close to, but not quite, 1. These observed values of k were several percent higher than we had originally calculated. This discrepancy was explained by Leo Szilard: in a water lattice, the fuel elements were so close together that fast neutrons from one rod could jump to an adjacent rod and cause fissions there. This interaction fast effect is, of course, well understood now, but, at the time, 45 years ago, it was regarded as a minor miracle.

Thus, water-moderated and water-cooled reactors were very much in the air in those earliest days, and on April 11, 1946, F. H. Murray and I issued a report, MONP-95, *High Pressure Water as a Heat Transfer Medium in Nuclear Power Plants*. But at that time our ideas about power reactors were dominated by our conviction that only the breeder made real sense; we hardly thought that nonbreeders, which simply burned enriched uranium, would be very important!

Karl Cohen was responsible for changing our perspective on this. He was the first to realize that gaseous diffusion, when fully rationalized, would produce enriched uranium at costs sufficiently low to allow its use in LRWs. Cohen occupied an all-but-unique position in those early days because he had participated in the development of the original theory of the diffusion cascade at Columbia University in 1942; and when Harold Urey was banished to Chicago by General Groves in 1943, Cohen also came to Chicago to work with Wigner on the design of heavy-water reactors. Cohen, therefore, was the first to command a detailed understanding of both reactors and diffusion plants. As matters turned out, slightly enriched uranium was eventually produced at a price that could be afforded in civilian reactors.

The third change was in the outlook for uranium ore. At the New Piles Committee, we spoke of 20,000 tons of uranium worldwide; by the late 1940s, we realized that uranium was much more plentiful—as it turns out, at least 1,000 times more plentiful—than we had estimated. Thus, the incentive for breeding was very much diminished—a situation that persists to this day.

Was the pressurized-water reactor chosen for commercial power because it was obviously the best choice? Not as I remember the matter. It was chosen for Shippingport after President Eisenhower had vetoed the Navy's proposal to build a nuclear aircraft carrier powered by a larger version of the *Nautilus* power plant. A demonstration of a power plant that would operate as part of an electrical utility was being urged by the Atomic Energy Commission. The only reactor that was on hand was the one designed for the cancelled aircraft carrier—what was more natural than to rescue Rickover's carrier reactor by putting it on land and operating it at Shippingport as part of the Duquesne Company's grid?

Given that LWR was chosen largely as a matter of expediency, was this a bad choice? At the time, Calder Hall, a Magnox, gas-cooled power reactor, was operating; the CANDU was in the final design stage; the Soviets had operated a small version of their RBMK; and two small prototypes aimed at breeding, Argonne's EBR-I and Oak Ridge's Aqueous Homogeneous Reactor, had operated. The Atomic Energy Commission hardly had a serious alternative to choose from, given its desire to *demonstrate* a peaceful use of nuclear power. To have diverted effort from the main light-water line would have incurred risks, and the arguments favoring the alternatives were not really compelling. After all, they too were expedient, not "best" choices: Calder Hall was an outgrowth of the Windscale plutonium plants, just as the Russian Obninsk was a variant of their plutonium producers; and CANDU, based on natural uranium, was an outgrowth of Canada's NRX experimental reactor.

The two primary aims of nuclear power—inexhaustible energy (that is, breeding) and economically competitive electricity—have both been demonstrated. The breeding ratio in France's sodium-cooled Phenix has been shown to be ~1.13; Admiral Rickover's thorium-based, ^{233}U seed-blanket light-water breeder has been shown to have a breeding ratio of around 1.01. These demonstrations of actual breeding have passed virtually unnoticed. I regard them as extremely important, since we now can say with certainty that nuclear fission, based on breeders that burn very low-grade ores, represents an all but inexhaustible source of energy. And, as I have already described, the goal of economically competitive nuclear electricity has been demonstrated in many places. What has not yet been demonstrated is the generation of electricity by a large-scale breeder that is cost-competitive today. The largest breeder, the 1200-MWe Super-Phenix, is too expensive, and the same probably can be said of the light-water breeder. On the other hand, if successors to Super-Phenix could be built more cheaply—a very likely prospect—or if the life of a breeder reactor could be extended to 100 or 150 years (instead of the planned 30–40 years), the cost of the electricity over the entire lifetime of the reactor could be competitive.

Super-Phenix has placed a cap on the cost of electricity for many future generations, if not for millennia, and this cap is surely less than twice current costs of electricity. Whether fusion or solar electricity will match this and will eventually displace the breeder remains to be seen. I therefore regard nuclear energy today as having all but achieved the primary goal we set for it at the New Piles Committee some 45 years ago—an inexhaustible energy source at a cost that ought not to escalate even as high-grade ore is exhausted.

The Need for a Second Nuclear Era

As one of the surviving participants of the New Piles Committee, I should feel very good; our main aim, expressed 45 years ago—power breeding—has been achieved! And competitive electricity from non-breeding light-water reactors, a goal that we hardly considered important at the time, is a reality in many places. But, despite these extraordinary successes, nuclear energy hangs in the balance and is all but dead in many countries. What went wrong, and how can we set it right?

Let me return to the New Piles Committee. Even then, Fermi seemed to sense that nuclear energy might encounter public opposition. I quote from Ohlinger's report of the April 26, 1944 meeting of the New Piles Committee, at which Fermi outlined his ideas for a fast breeder that fed its fissile plutonium to small satellite power plants: "There may be nontechnical objections to this arrangement, for example, the shipment of 49* to smaller consuming plants offers the serious hazard of its falling into the wrong hands." And I can remember as if it were yesterday (although I cannot document it) Fermi's pronouncement at one of these meetings that, for the first time, mankind would be confronted with enormous amounts of radioactivity; we must not assume that this will be accepted easily by society, he warned.

Fermi's warning is proving to be closer to the mark than the optimistic predictions of early enthusiasts, including me. One explanation for this turn of events is that the dawning of the nuclear age has coincided with the dawning of what I call the "age of anxiety." Although life expectancy in most of the developed world has increased enormously since the turn of the century, we are much more anxious about survival—as individuals, as a society, and even as organisms on an inhabitable Earth—than ever before. We worry about low levels of chemical insult, about the possibility of nuclear obliteration, about irreversible pollution of the planet. All of this is greatly exacerbated by television and by new trends toward participatory democracy. Some of our worries are justified: nuclear war, possibly the greenhouse effect, the possibility of a catastrophic accident like those at Chernobyl and Bhopal. Some are without scientific justification, such as the exaggerated claims that radiation (or other hazards, including toxins) at levels much lower than background pose any hazard to health. But, in the age of anxiety, the public does not distinguish between those worries that are real and those that are unjustified.

In the case of nuclear energy, we must distinguish among the public's primary concerns. First, there is *proliferation of nuclear weapons*. I would judge that, until the Iraq war, this was more of a worry for the professional arms-control expert than for the public at large. The Iraq war may have changed the public's perception, but I am unable to document this.

The second worry is radioactive waste disposal. We must remember that James Conant, President Roosevelt's personal monitor of the Manhattan Project, predicted in 1953 that nuclear power would not be worth the candle because waste disposal would prove to be intractable.

---

*49 was the wartime code name for ^{239}Pu.

The public is by and large convinced that Conant was right: wastes are the public's main concern about nuclear energy. Yet I would insist that wastes fall largely into the category of *un*justified concerns. The reason is that, once the spent fuel has been removed from the reactor and has been dispersed in a cooling pond or in a disposal cask, there is no longer enough energy being generated to cause widespread dispersal of *large* amounts of radioactivity. No analysis of a high-level depository, including the accompanying transport system, has identified a credible accident that disperses really large amounts of radioactivity over really large amounts of land. An accident that imposes lethal doses on a *few* people in the immediate vicinity of a depository—yes, although with small probability; or an accident that imposes very *low* doses on *large* numbers of people—also possible. But an accident, such as Chernobyl, that imposed large doses on large groups of people—no. And if, as I believe, the weight of evidence supports the view that doses of x rays at less than background are hardly harmful and may even be beneficial, I would insist that the public's concern over wastes is much exaggerated. It is merely a manifestation of our fearfulness in this age of anxiety.

Not so with the possibility of a reactor accident. As Chernobyl showed, and as Norman Rasmussen had estimated in 1975, an uncontained meltdown of a large nuclear reactor could be a catastrophe of immense proportion. Unlike waste disposal, it can impose large doses on large numbers of people as well as contaminating large tracts of land. The nuclear community has tried to deal with this reality by invoking probabilistic arguments: the *a priori* probability is very small—say, less than 10^{-6} per reactor year, or even lower—for accidents as large as the one at Chernobyl. But I don't think the public accepts such probabilistic arguments, even when the numbers show the absolute dangers are comparable to the dangers accepted in other, more familiar technologies, such as hydroelectric dams. What the public seems to want is a return to the time when an engineer would be able to say "This is safe—period," not "This is relatively safe," or the "Probability of an accident that causes unacceptable consequences is a small number, roughly 10^{-6} per reactor-year."

Can we develop nuclear reactors whose safety is deterministic, not probabilistic, and that, if developed, would meet the public's yearning for an assurance of safety, not simply an assurance about the probability of safety? This is the task that has engaged many nuclear engineers in a search, now 10 years old, for an inherently safe reactor. Now, in some sense, a device that produces 200 megawatts of afterheat and is immune from meltdown under every circumstance, foreseeable and unforeseeable, is a contradiction in terms. But a device whose safety depends on the working of immutable laws of nature, with a minimum of interventions either mechanical, electrical, or human—a device, in short, whose safety is so *transparent* that the skeptical elite, as well as the informed public, will regard it as safe—this I regard as eminently possible and worthwhile. There are now at least two ideas for reactors that are largely inherently safe—the SECURE-P (also known as PIUS) of the ASEA-ABB company of Switzerland and Sweden; and the Modular HTGRs of General Atomic Company in the United States and of the KWU Company in Germany. Argonne National Laboratory is developing a metallic-fueled small breeder that embodies almost as many passively safe features as do these two. The advanced developments sponsored by Westinghouse and General Electric aim at producing relatively small reactors that are incremental improvements over existing LWRs yet provide substantially more passive safety than existing reactors. In Canada, the research reactor SLOWPOKE and a power version of SLOWPOKE provide important elements of passive safety.

I cannot say where the quest for inherently safe (or at least transparently and passively safe) reactors will end. Will it lead to a second nuclear era, risen from the ashes of Chernobyl and Three Mile Island, in which the public accepts reactors as safe, not because quantitative arguments predict a very low probability of an accident with unacceptable consequences but because the safety of these reactors are understandable and plausible? Or will the second nuclear era dwindle in a chaos of recrimination and protest because the public simply cannot be convinced, even by reactors that embody passive and inherent safety?

What the public's reaction will be surely depends on the alternatives to, and upon the incentives for, nuclear power in the next 50 years. As for incentives, the greenhouse effect may be of great importance. I have estimated that, to stabilize carbon dioxide in the atmosphere through nuclear energy, the world might have to build as many as 5,000 very large reactors, producing about 300 quads of primary energy. A 1-GWe LWR without recycling burns 150 tons of natural uranium per year; with recycling, 80 tons/year; whereas a breeder burns only three tons per year. Thus, the yearly demand for uranium in a world that derives 300 quads from nuclear power would be 750,000 tons of uranium for a nonrecycling LWR, 400,000 tons of uranium for a recycling LWR, and 15,000 tons of uranium for breeders. Since the "speculative" reserve of uranium at less than \$130 per kilogram is around 20×10^6 tons, this resource could sustain a 5,000-GWe nuclear deployment for about 1,000 years if based on breeders, but only 50 years if based on a recycling LWR, and 27 years if based on a nonrecycling LWR. I do not regard these estimates as reliable, especially since we do not know the actual size of the world's uranium reserve. The estimates do suggest that control of emissions of carbon dioxide may be the strongest motivation for continued vigorous development of the breeder.

Even if the uranium supply is adequate, a 5,000-reactor world is impossible if the core-melt probability is as high as 10^{-4} per reactor-year, or even 10^{-5} per reactor-year, since this leads to a core melt every 2 to 20 years. And as Chernobyl showed, a single core melt of that magnitude puts an end to nuclear power in many countries. The public may regard greenhouse warming as worse than Chernobyl, and they may not demand inherent safety as the price of a second nuclear era. I would not count on this, however; I believe the nuclear community, descendants in a way of Fermi and the New Pile Committee, have no choice but to pursue ideas for passively safe reactors.

For in a way there are alternatives to nuclear energy. There was no heaven-ordained requirement that the age of fossil fuel be replaced by the age of fission, nor a cosmically ordained anthropic principle that required the number of neutrons emitted per neutron absorbed in ^{235}U to exceed 1 so that a chain reaction be possible, or exceed 2 so that breeding is possible. Had fission not been available, we would, willy-nilly, be conserving energy at a much faster rate than now, we would be pushing fusion even harder, and, as a last resort, we would turn to solar energy. A solar world in which primary energy is three times as expensive as it is today is hardly an impossible world, especially since, in such a world, energy would be much more strongly conserved than it is now. We would not be relegated to a Malthusian poverty, were the only reactor upon which we could depend for energy located 150×10^6 kilometers away from the Earth.

But Hahn and Strassmann's discovery of 50 years ago and God's providence in adjusting the nuclear constants so as to make a power breeder practical have given us another option. We nuclear engineers of the first nuclear era have had success, yes, with our 500 commercial reactors and our practical breeders. But the job is only half finished. The generation that follows us must resolve the profound technical and social questions that are convulsing nuclear energy. The challenge is clear, even the technical paths to meet the challenge are clear. All of us old-timers wish we could be here to see how these challenges will be met; but, even if we are not here, we wish the new generation well in fashioning an acceptable second nuclear era.

Index